Saas-Fee Advanced Course 29
Lecture Notes 1999

Springer
Berlin
Heidelberg
New York
Barcelona
Hong Kong
London
Milan
Paris
Tokyo

F. Palla H. Zinnecker

Physics of Star Formation in Galaxies

Saas-Fee Advanced Course 29
Lecture Notes 1999
Swiss Society for Astrophysics and Astronomy

Edited by A. Maeder and G. Meynet

With an Introduction by George Herbig

With 204 Figures and 25 Tables

Springer

Francesco Palla
Osservatorio Astrofisico di Arcetri, Università di Firenze
Largo Enrico Fermi, 5, 50125 Firenze, Italy

Hans Zinnecker
Astrophysikalisches Institut Potsdam, Universität Potsdam
An der Sternwarte 16, 14482 Potsdam, Germany

Volume Editors:

André Maeder
Georges Meynet
Observatoire de Genève
Ch. des Maillettes 51
CH-1290 Sauverny, Switzerland

This series is edited on behalf of the Swiss Society for Astrophysics and Astronomy:

Société Suisse d'Astrophysique et d'Astronomie
Observatoire de Genève, ch. des Maillettes 51, 1290 Sauverny, Switzerland

Cover picture: Hubble Space Telescope image of NGC 604, a giant star forming region, 1500 light years across, in the nearby spiral galaxy M33.

Die Deutsche Bibliothek – CIP-Einheitsaufnahme

Physics of star formation in galaxies: with 25 tables
Saas Fee Advanced Course 29. F. Palla; H. Zinnecker. Swiss Society for Astrophysics and Astronomy.
Ed. by A. Maeder and G. Meynet. –
Berlin; Heidelberg; New York, Barcelona; Hong Kong; London; Milan; Paris; Tokyo: Springer, 2002
(Saas Fee Advanced Courses; 29. 1999)
(Physics and astronomy online library)
ISBN 3-540-43102-0

ISBN 3-540-43102-0 Springer-Verlag Berlin Heidelberg New York

Springer-Verlag Berlin Heidelberg New York
a member of BertelsmannSpringer Science+Business Media GmbH

http://www.springer.de

Typesetting: Camera-ready copy from authors/editors
Cover design: *design & production* GmbH, Heidelberg
Printed on acid-free paper SPIN 10723707 55/3141/XO - 5 4 3 2 1 0

Preface

In a letter of December 10, 1692 to the young minister Richard Bentley, Isaac Newton wrote *"... the matter on ye outside of this space would by its gravity tend towards all ye matter on the inside & by consequence fall down to ye middle of the whole space & there compose one great spherical mass. But if the matter was eavenly diffused through an infinite space, it would never convene into one mass but some of it convene into one mass & some into another so as to make an infinite number of great masses scattered at great distances from one to another throughout all ye infinite space. And thus might ye Sun and Fixt stars be formed supposing the matter were of a lucid nature."* These magnificent lines, which propose a cosmological answer to the question why gravitation is not packing all the matter into one single object in the universe, represent the first study on the growth of density fluctuations. With a profound inspiration, they enlighten for the first time the dominant role of gravitation in the formation of stars.

In this course on the "Physics of Star Formation in Galaxies", Francesco Palla and Hans Zinnecker are successfully studying the problem set by Newton, namely how the interstellar matter may "convene into one mass and some into another". They do it with a high quality requirement and elegance, in a way meritorious of their eminent ancestor in the field. Francesco Palla covers, with great clarity and a deep physical insight, the theoretical aspects of star formation, a subject in which he has himself made major contributions, in particular for our understanding of the pre-Main Sequence evolution with mass accretion. A unique information, never provided in literature, is given by Hans Zinnecker, who shows us how essential observations can be performed with the VLT, the HST, ISO, the Keck telecope, the ESO 3.6 m telescope and the CHANDRA X-ray observatory on various aspects of star formation, on which he is a master: protostars, young binaries, jets, disks, low mass objects, etc...

The 29th Course of the Swiss Society for Astrophysics and Astronomy was held with 72 participants in Les Diablerets, a nice mountain resort of the Swiss Alps from 22 to 29 March 1999. The lecturers were Francesco Palla, Hans Zinnecker and Douglas Lin. From the promised manuscripts, two were received. George Herbig gave to Hans Zinnecker an historical introduction for this volume "Star Formation: the Early History". We see that questions, which today have an evident answer for all students, such as what is the reservoir of matter making the stars, needed decades of efforts to be clarified. We thank George Herbig very much for this introduction, which provides us with some generally unknown parts of the History of Sciences.

Geneva,
November 2001

André Maeder
Georges Meynet

Star formation lecturers: F. Palla, D. Lin, H. Zinnecker (*from left to right*). (Photo by A. Maeder)

Contents

Historical Introduction. Star Formation: the Early History
George Herbig 1

Pre-Main-Sequence Evolution of Stars and Young Clusters
Francesco Palla 9

I. Initial Conditions of Pre-Main-Sequence Evolution 9
1 The Classical Theory of Pre-Main-Sequence Evolution 9
1.1 Basic Results 9
2 The Impact of Star Formation 12
2.1 Protostellar Collapse: the Mass Accretion Rate 13
2.2 Classical Collapse Models 14
2.3 More Realistic Models 17
3 Protostellar Evolution 20
3.1 Central D-Burning 22
3.2 Radiative Barrier 23
3.3 The Onset of D-Shell Burning and Gravitational Contraction . 25
3.4 The Stellar Birthline 26
3.5 Effects of the Accretion Flow 26
II. Pre-Main-Sequence Evolution 32
4 Evolutionary Models 32
4.1 Protostellar Initial Conditions 32
4.2 Stellar Structure Equations 35
4.3 Equation of State 37
4.4 Opacity 38
4.5 Treatment of Convection 40
4.6 Nuclear Reactions 40
5 The H-R Diagram 41
5.1 Low-Mass Stars: $0.1 \lesssim M_*/M_\odot \lesssim 2$ 43
5.2 Intermediate-Mass Stars: $2 \lesssim M_*/M_\odot \lesssim 10$ 46
5.3 Massive Stars: $M_* \gtrsim 10 M_\odot$ 48
6 Tests to PMS Evolutionary Diagrams 51
6.1 Comparison of Tracks 51
6.2 The Low-Mass M Dwarfs and the End of the Main Sequence . 54

6.3 Derivation of Observable Quantities from Evolutionary Tracks 57
6.4 Pre-Main-Sequence Binaries 60
III. Physical Processes in PMS Stars 70
7 Lithium Depletion in Young Stars 70
7.1 Rotation and Depletion 71
7.2 Spectroscopic Classification Based on Lithium 72
7.3 Lithium and Brown Dwarfs 72
8 Mass Accretion in Young Stars 74
8.1 Disk Accretion 74
8.2 Effects on Evolutionary Tracks 76
9 Rotation and PMS Evolution 80
9.1 Rotational Properties of Young Stars 80
9.2 Structural Effects 85
10 Pulsational Instabilities in PMS Stars 87
10.1 The Input Models 88
10.2 The Instability Strip 89
IV. PMS Evolution of Clusters and Associations 97
11 High Mass Star Formation 98
11.1 The Search for Massive Protostars 99
11.2 The Evolution of Ultracompact HII Regions 103
12 Clustering of Young Stars 105
12.1 Clustering and Mass Segregation 108
12.2 Effects on the Environment 109
12.3 Clustering and Binary Formation 112
12.4 Clustering and Star Formation Efficiency 113
12.5 Clustering and the IMF 115
13 Stellar Populations in Clusters and Associations 117
13.1 H-R Diagrams 118
13.2 Star Formation Histories 124
13.3 The Post-T Tauri Problem 128

Observations of Young Stellar Objects (YSO)
Hans Zinnecker 135

I. Introduction 135
1 How to Find Young Stars and Protostars 136
II. Disks, Jets and Protostars 140
2 A VLT/ISAAC Study of Circumstellar Disks and Envelopes Around YSOs 140
2.1 Abstract 140
2.2 Scientific Rationale 140
2.3 Immediate Objective 142
2.4 Telescope Justification 143
2.5 Results 143

3 Infrared H_2 Imaging of the Highly Symmetric Proto-stellar Jets HH 212 and HH 111 146
3.1 Abstract 146
3.2 Scientific Background: Star Formation and Herbig-Haro Jets . 147
3.3 Scientific Goals 150
3.4 The Need for HST 152
3.5 Results 153
4 An Unbiased mm Continuum Survey for Protostellar Outflow Sources in the Orion A Cloud 155
4.1 Abstract 155
4.2 The Need for an Unbiased Millimetre Survey for Protostellar Outflow Sources 155
4.3 Framework: A Survey for Jets and Protostars in L 1641 157
4.4 Technical Details 159
4.5 Results 159
5 Mid-Infrared Imaging of Deeply Embedded Protostellar Sources in the Orion A Cloud 161
5.1 Abstract 161
5.2 Scientific Justification 161
5.3 Why Do We Need Mid-Infrared Observations at Keck? 162
5.4 Technical Remarks 163
5.5 Results 164
6 A Study of Faint Disks Around X-ray Selected Young Low-Mass Stars in an OB Association 166
6.1 Abstract 166
6.2 Description of the Proposed Programme 167
6.3 Results 169
III. Young Binaries 172
7 Statistics of Young Binary Stars: Implications for (Binary) Star Formation 172
7.1 Young Binary Stars: Why Care? 172
7.2 Observational Methods 172
7.3 Results 175
7.4 Implications 177
7.5 Future Prospects 180
7.6 Discussion 181
8 Low-Mass Binary Companions to Intermediate-Mass Stars in the Sco-Cen OB Association 184
8.1 Abstract 184
8.2 Description of the Proposed Programme 184
8.3 Results 187
9 Multiplicity of Massive Stars in the Orion Nebula Cluster and Implications on Their Formation Mechanism 189
9.1 Abstract 189

9.2 Introduction 189
9.3 Observations 189
9.4 Results 190
9.5 Conclusions 195

IV. Low Mass Stellar Content of Young Star Clusters and Associations 197

10 The Low-Mass Initial Mass Function in the Upper Scorpius OB Association 197
10.1 Abstract 197
10.2 Scientific Justification 197
10.3 Results 201

11 The Full Population of T Tauri Stars of the Young Stellar Cluster IC 348 202
11.1 Abstract 202
11.2 Scientific Justification 202
11.3 Technical Feasibility 206
11.4 Results 206

12 A Study of the Low-Mass Stellar Population in the Galactic Starburst Region NGC 3603 208
12.1 Abstract 208
12.2 Description of the Proposed Programme 208
12.3 Required Observing Time 211
12.4 Results 213

13 The Low-Mass Pre-MS Stellar Content of the 30 Dor Starburst Cluster 215
13.1 Abstract 215
13.2 Scientific Background 215
13.3 Previous Work 217
13.4 Summary of the Scientific Goals and Method 218
13.5 The Need for NICMOS/HST 218
13.6 Description of the Observations 219
13.7 Results 220

V. Appendix: The Suite of Large Ground-Based Telescopes 224

Subject Index 227

List of Previous Saas-Fee Advanced Courses

!! 1999 Physics of Star Formation in Galaxies
F. Palla, H. Zinnecker

!! 1998 Star Clusters
B.W. Carney, W.E. Harris

!! 1997 Computational Methods for Astrophysical Fluid Flow
R.J. LeVeque, D. Mihalas, E.A. Dorfi, E. Müller

!! 1996 Galaxies Interactions and Induced Star Formation
R.C. Kennicutt, F. Schweizer, J.E. Barnes

!! 1995 Stellar Remnants
S.D. Kawaler, I. Novikov, G. Srinivasan

* 1994 Plasma Astrophysics
J.G. Kirk, D.B. Melrose, E.R. Priest

* 1993 The Deep Universe
A.R. Sandage, R.G. Kron, M.S. Longair

* 1992 Interacting Binaries
S.N. Shore, M. Livio, E.J.P. van den Heuvel

* 1991 The Galactic Interstellar Medium
W.B. Burton, B.G. Elmegreen, R. Genzel

* 1990 Active Galactic Nuclei
R. Blandford, H. Netzer, L. Woltjer

! 1989 The Milky Way as a Galaxy
G. Gilmore, I. King, P. van der Kruit

! 1988 Radiation in Moving Gaseous Media
H. Frisch, R.P. Kudritzki, H.W. Yorke

! 1987 Large Scale Structures in the Universe
A.C. Fabian, M. Geller, A. Szalay

! 1986 Nucleosynthesis and Chemical Evolution
J. Audouze, C. Chiosi, S.E. Woosley

! 1985 High Resolution in Astronomy
R.S. Booth, J.W. Brault, A. Labeyrie

! 1984 Planets, Their Origin, Interior and Atmosphere
D. Gautier, W.B. Hubbard, H. Reeves

! 1983 Astrophysical Processes in Upper Main Sequence Stars
A.N. Cox, S. Vauclair, J.P. Zahn

* 1982 Morphology and Dynamics of Galaxies
J. Binney, J. Kormendy, S.D.M. White

! 1981 Activity and Outer Atmospheres of the Sun and Stars
F. Praderie, D.S. Spicer, G.L. Withbroe

* 1980 Star Formation
J. Appenzeller, J. Lequeux, J. Silk

* 1979 Extragalactic High Energy Physics
F. Pacini, C. Ryter, P.A. Strittmatter

* 1978 Observational Cosmology
J.E. Gunn, M.S. Longair, M.J. Rees

* 1977 Advanced Stages in Stellar Evolution
I. Iben Jr., A. Renzini, D.N. Schramm

* 1976 Galaxies
K. Freeman, R.C. Larson, B. Tinsley

* 1975 Atomic and Molecular Processes in Astrophysics
A. Dalgarno, F. Masnou-Seeuws, R.V.P. McWhirter

* 1974 Magnetohydrodynamics
L. Mestel, N.O. Weiss

* 1973 Dynamical Structure and Evolution of Stellar Systems
G. Contopoulos, M. Hénon, D. Lynden-Bell

* 1972 Interstellar Matter
N.C. Wickramasinghe, F.D. Kahn, P.G. Metzger

* 1971 Theory of the Stellar Atmospheres
D. Mihalas, B. Pagel, P. Souffrin

---

* Out of print

! May be ordered from Geneva Observatory
Saas-Fee Courses
Geneva Observatory
CH-1290 Sauverny
Switzerland

!! May be ordered from Springer Verlag

Historical Introduction.
Star Formation: the Early History

George Herbig

As evident from its colloquial style, this essay was never intended for publication; it was prepared for a seminar on star formation given at the IfA in March 2000. Others who participated in the early development of the subject may have seen it otherwise, but this is how it looked to me. – G.H.

It is not easy to reconstruct what the astronomical community thought of star formation prior to about 1930. In 1854–1863 Helmholtz and Kelvin (the original references are given by Chandrasekhar [6] p. 484) had considered the possibility that the Sun's luminosity was derived from slow contraction, but the time scale was very short: contraction from infinite radius to the present value, at the current luminosity, could have sufficed for only 10 to 30 Myr. The reigning idea thereafter seemed to be that stars lose mass by radiating it away, and hence that a star's mass decreases systematically with age. Stars were supposed to begin their lives as massive red giants, evolve up the giant branch to the main sequence at type B, and then ever more slowly down the main sequence toward the M dwarfs. I can find no explanation of where those massive red giants came from. One senses the attitude that the origin of stars was something unknown and unknowable: it happened in the remote past under circumstances that we were unlikely to fathom.

Origins aside, after the equivalence of mass and energy was recognized, the numerical consequences of radiation via mass loss could be calculated. This led to serious conflict with observation: in this picture, the ages of stars of different mass had to be grossly different, so the existence of both high- and low-mass stars in the same cluster, all presumably formed at the same time, could not be explained. Furthermore, if luminosity was the product of complete annihilation of matter, the time scale was very long: it was estimated that it would require about 6 (12) (*i.e.* 6×10^{12}) years for the Sun to burn itself down its present M, L and R.

The discovery of the thermonuclear conversion of H to He as the energy source for main sequence stars is credited to Bethe, in 1938, and indeed he did work out the detailed nuclear physics of the process. But the idea was not new. Eddington [7] had remarked that "an age of 1300 million years has been assigned to the oldest sedimentary rocks", and since "the sun must be much older ... it is now generally agreed that the main source of a star's energy is subatomic". He then considered, among other ideas, the reasons for considering "the transmutation of hydrogen into helium or higher elements as the main source of stellar energy". The physics of 1930 could not do much more, but by 1937 things had moved on. In Chandrasekhar's book [6] he describes the papers of Gamow, von Weiszäcker and others that culminated in Bethe and the situation we know today: see Bethe's own account [2].

Payne had demonstrated the high abundance of hydrogen in stellar atmospheres, which made it a plausible and abundant source of energy. But it became clear that the H-burning lifetimes of the massive O stars, even with all that H to consume, were so short that the issue of their formation under present-day conditions could not be avoided: the excuse that they must have formed under mysterious circumstances in the remote past was no longer tenable. For example an O6 V radiates about 1.4 (39) erg sec^{-1}, and has a mass of about 40 $M_{\odot}$, or 8.0 (34) gm. Since the conversion of H into He liberates 6 (18) erg gm^{-1}, even if the whole mass of H in that star were burnable, it would last only about 10 Myr; in fact only about 10 % of the mass is consumable, so the life expectance of that star on the main sequence is really only about 1 Myr.

If so, then where could they be born: what reservoir of adequate mass existed? The formation of such stars must be going on somewhere nearby, since there is little time for them to move very far from their birthplaces. The problem was specially acute because massive stars are often found in clusters containing many less massive stars, totalling hundreds of solar masses, obviously all formed somewhere at about the same time. So there ought to be young stars of lesser mass, perhaps just like the Sun, formed very recently.

The only known reservoir of mass in those amounts was of course the cloudy interstellar medium, and indeed most OB stars are found in or not far from dense collections of interstellar material. But there were competing ideas. Hoyle and Lyttleton defended the view that old stars could accrete sufficient mass by accretion during their passage through interstellar clouds to rejuvenate them. If so, H-burning in isolation would give only a lower limit on the ages. The objections to this idea were that accretion is not that easy for hot stars, and such stars do not have surface convection zones, so material collected on the surface would remain there.

Ambartsumian, contemplating the apparent movement of OB stars out of a small volume as implied by Blaauw's work on expanding associations (more on that later), pushed the idea that stars might be formed by the disintegration of dark, unstable super-massive objects of which we know nothing, rather like the fission of a massive nucleus.

The whole issue had been around of a much longer time but in another form: the question of the origin of the solar system, which had been debated at least since the time of Laplace. There were two points of view: either 1) the Sun had been formed somewhere as a single star, and then afterward encountered another star which, passing nearby, pulled off material some of which formed the planets. Or 2), the Sun and planetary system were formed in a single process, which is the present point of view. In either case, that event must have happened somewhere else, a long time ago, and I have not seen any remarks by those earlier speculators on how the Sun itself may have come about. The idea that some stages of the process of the formation of Sun-like stars might actually be observable, and indeed ought to be going on at least in those young clusters containing young OB stars, apparently never occurred to anyone until the evidence forced itself upon our attention.

Our present ideas on star formation gradually sank into the consciousness of the community after WW II, but it took some time. The observational breakthrough, although there were no press releases or publicity at the time, and none of the influential theorists of the day apparently noticed it, was Joy's discovery of the T Tauri 'variables' [18] [19].

The sequence of subsequent events, and the dawning of enlightenment, went about like this:

1946: Whipple [28], recognizing that "the interstellar matter provides the only obvious source of stellar building material", suggested that star formation could begin when a parcel of interstellar dust, exposed from all sides to the galactic radiation field, would be squeezed and so forced into contraction; his estimate of the duration of this process for 1 solar mass was 1.1 Gyr.

1948: possibly as a result of the interest of Whipple and Bok, a theme of the Harvard Centennial Symposium in that year was that "in some cosmic clouds we are now witnessing the operation of the process of star formation". Bok & Reilly [5] had already described 'globules', small dark spots seen in silhouette against some bright nebulae. In that Symposium, Bok [4] discussed the contraction of such an object under the radiation pressure of the galactic field, and its increase in mass by accretion of interstellar material from the neighborhood, and implied that such globules would gradually collapse "to become star-like objects". Note that Bok, like Whipple and Spitzer (at about that same time) believed that this did not happen inside a cold, dark molecular cloud as we believe today, but in the full glare of the interstellar radiation field, with radiation pressure as the initiating agent.

In that same volume, Spitzer (p. 87) noted that the "similiarity of distribution between [high luminosity] stars and interstellar matter suggests that such stars have been formed from interstellar matter, and may even be forming at the present time", but although "the results presented here are therefore favorable to the hypothesis that [high luminosity] stars, and probably also some stars of lesser mass, have originated recently from interstellar matter, this hypothesis remains extremely tentative at the present time". In a later review [25] he said much the same: "young stars are present in and only in those regions of space where interstellar matter is also found. The circumstantial evidence for the generation of stars from clouds seems very strong".

Spitzer may have been the first to use a term we hear often nowadays: during "gravitational contraction an aggregation of dust and atoms will be stable against most disruptive influences; it should no longer be regarded as a cloud, and will here be called a "protostar" ".

Whipple was interested primarily in the formation of the planetary system, and his Symposium contribution (p.109) sounds surprisingly modern: he considered the collapse of a cloud of solar mass and initial radius of 30,000 AU, and proposed that sub-clouds in the original larger cloud "spiraled inward by accretion at the proper rate to be left in approximately the orbits of the present planets when the main cloud underwent its final rapid collapse" to form the Sun. *I.e.* he thought that a planetary system could form as part of the process that produced the central star.

These were the first suggestions as to the specific processes that could lead to the formation of a star, but except for Bok's globules (which today we do not regard as the object which itself undergoes gravitational collapse, although some globules may contain pre-main sequence stars within them) there was no association of these ideas with anything observable in the sky. None of these people mentioned Joy's T Tauri stars (TTS), nor did Spitzer here or (as far as I know) in any of his later writings on the subject. The link had to come from the observers.

1947 - 1954: real credit must be given here to the Russian astronomers Ambartsumian, Kholopov and Parenago who in those years argued that the TTS stars were young stars of intermediate mass, still in process of gravitational contraction, exactly what we believe today. But their first papers were published either in Russian, or in places where most western astronomers were unlikely to pay attention, so their ideas were largely unheeded. References to some of those papers are given by Herbig [16]. There is a detailed account of Ambartsumian's work by Mirzoyan [21].

1950: in the west, it was Struve who picked up and pursued Joy's lead: Struve & Rudkjøbing [26] obtained spectra of a number of faint stars projected upon the dense obscuration near ρ Oph, and found that several were late-type dwarfs with emission lines just like Joy's TTS. He had accepted Ambartsumian's arguments that associations containing early-type stars must be young and have their origin in the surrounding interstellar material. Struve was one of the few westerners who could and did read the Russian originals, but he was then (and a year later in his book *Stellar Evolution*) still influenced by the Hoyle/Lyttleton idea that these stars might be interlopers from the field, and that the emission spectra were powered by the infall energy of accreting cloud material.

There was still no hint that these objects might be the very things that western theorists were talking about. Still more TTS stars were found by Joy[20], by Haro [8] [9] [10] in Sco-Oph and in Orion, and in the Orion Nebula by Herbig [12], but without any inspired speculation as to their significance. No firm conclusion had been reached even two years later:

1952: Payne-Gaposchkin [22], despite the provocative title of her book (*Stars in the Making*), was still not sure: "perhaps [the peculiar variable stars within the Orion Nebula] are actually being formed within it" or perhaps they are pre-existing stars that have entered and are accreting.

More fence-straddling: Herbig [13]: "the question arises if these [TTS] are ordinary field objects that have moved into the nebulae and exhibit emission spectra and light variations as a result of some interaction process with the nebular material, or whether they are young stars that have been formed within the clouds. The Russian astronomers V. Ambartsumian and P. Kholopov ... believe that the T-associations are composed of newly-formed main-sequence stars, which in the early stages of their lives behave as unstable objects with irregular variations in light ... [the] evidence is somewhat ambiguous". But in that same year:

Blaauw [3] published his work on expanding associations of OB stars, which suggested that these young stars were moving outward from their birthsites on a time scale of 1 - 2 Myr. And in:

1954: Herbig [14] discovered the TTS in IC 348, a cluster that, according to Blaauw, participates in the expansion of the Per OB2 association. This seemed direct demonstration that TTS were young, and indeed were examples of those low-mass stars that were expected to accompany the formation of high-luminosity stars from interstellar clouds. But more tip-toeing around: "did the event that produced the high luminosity members of the ζ Per group also result ... in the formation of stars of low luminosity ... "? Then: it is "... probably correct", but then, wringing hands: "The present results do not by themselves imply that all T Tauri-like objects are of recent formation". But surely if they were formed at the same time as the OB stars of the association with an expansion time of a few Myr, at least those TTS in IC 348 must be young!

1954 - 1955: Theorists began to appreciate that young stars could be recognized by their location in the H-R diagram. Salpeter was apparently the first clearly to say so, initially at a symposium at Michigan in 1953, then again in the 1953 Liège symposium [23], where he recommended an "attempt to study whether the low luminosity part of the main sequence is actually missing in young star association[s], to look for reddened stars to the right of the main sequence and to look for similar effects in gas and dust clouds where star formation is still suspected to go on". The following year Henyey, LeLevier, & Levée [11] actually calculated pre-main sequence radiative tracks for contracting stars of masses between 0.65 and 2.291 $M_{\odot}$.

None of these people knew or commented on the fact that the TTS, which had then been known for about a decade met these specifications, or that Joy [20] had already remarked on the odd fact that the K- and M-type TTS he had found in the Taurus clouds were much too bright for their main sequence counterparts: exactly what Salpeter in 1953 was urging observers to look for.

1955: Herbig [15]: "The central question is: are the TTS variables new stars, recently formed or still forming within the nebulae, or are they ordinary field stars that have encountered the gas and dust clouds accidentally, and are in process of being modified ... by their environment? ... Three years ago ... I concluded that it was 'not possible to make a clear-cut and entirely acceptable decision between [these] two opposing alternatives ... Today, in my opinion, we are able to take a more positive stand. I believe that the evidence now available favors the hypothesis that the TTS as a class are new objects, genetically associated with the clouds in which they are found".

1956: Walker's paper [27] containing the photoelectric color-magnitude diagram of the young cluster NGC 2264 pulled all this together, showing as it did the elevated main sequence where the TTS were located. Sandage [24] gives a very good review of how the situation looked in 1957.

1958: Baade is not reported as having said so directly (in the printed version of his Harvard lectures [1]), but clearly he believed that the TTS in Orion and NGC 2264 have recently formed there, that star formation is on-going, and that star formation in such a region continues over a significant time interval. From that point on, the question seemed settled to the satisfaction of most people in the field. Much of this history is summarized in Herbig (1970).

After the glacial advances of these painful decades, the subject underwent a drastic revolution, with the advent of non-photographic detectors, exploitation of the infrared, and radio-frequency spectroscopy/ imaging of molecular clouds and their contents.

References

1. W. Baade: Evolution of Stars and Galaxies (ed. C. Payne-Gaposchkin), (Harvard Univ.) (1963)
2. H. A. Bethe: Physics Today, Sept. 1968, p.36 (1968)
3. A. Blaauw: BullAstrInst Netherlands **11**, 405 (1952)
4. B. J. Bok: Centennial Symposia (Harvard Obs. Monographs no. 7), p.53 (1948)
5. B. J. Bok & E. F. Reilly: ApJ **105**, 255 (1947)
6. S. Chandrasekhar: An Introduction to the Study of Stellar Structure (Chicago) (1939)
7. A.S. Eddington: The Internal Constitution of the Stars, (Cambridge Univ. Press), ch. xi (1930)
8. G. Haro: AJ **54**, 188 (1949)
9. G. Haro: AJ **55**, 72 (1950)
10. G. Haro: ApJ. **117**, 73 (1953)
11. L. G. Henyey, LeLevier, & Levée: PASP **67**, 154 (1955)
12. G.H. Herbig: ApJ **111**, 15 (1950)
13. G.H. Herbig: JRASC **46**, 222 (1952)
14. G.H. Herbig: PASP **66**, 19 (1954)
15. G.H. Herbig: Non-Stable Stars (IAU Symposium 3), p.3 (1957)
16. G.H. Herbig: AdvAstrAp **1**, 47 (1962)
17. G.H. Herbig: Observational Astrophysics (Berkeley: U. Calif.), p.237 (1970)
18. A.H. Joy: PASP **54**, 17 (1942)
19. A.H. Joy: ApJ **102**, 168 (1945)

20. A. H. Joy: ApJ **110**, 424 (1949)
21. L. V. Mirzoyan: Problems of Physics and the Evolution of the Universe (Yerevan: Armenian Acad. Sci.), p. 209 (1978)
22. C. H. Payne-Gaposchkin: Stars in the Making (Cambridge: Harvard), p. 27 (1952)
23. E. E. Salpeter: Mem.Soc.R.Sci. Liège 14 (Les Processus Nucléaires dans les Astres), p. 116 (1954)
24. A. Sandage: Stellar Populationsxi (Vatican)*, p. 149 (1958)
25. L. Spitzer: J.Wash.Acad.Sci. **41**, 309 (1951)
26. O. Struve & M. Rudkjøbing: ApJ **109**, 92 (1950)
27. M. F. Walker: ApJSuppl **2**, 365 (1956)
28. F. L. Whipple: ApJ **104**, 1 (1946)

* This volume has the formal title: Semaine d'Etude sur le Probleme des Populations Stellaires, Scripta Varia 16, Pontifical Academy of Sciences (Vatican City)

Pre-Main-Sequence Evolution of Stars and Young Clusters

Francesco Palla

Osservatorio Astrofisico di Arcetri
Largo Enrico Fermi, 5
50125 Firenze, Italy

I. Initial Conditions of Pre-Main-Sequence Evolution

1 The Classical Theory of Pre-Main-Sequence Evolution

The location of optically visible Pre-Main-Sequence (PMS) stars in the H-R diagram combined with the use of theoretical evolutionary tracks have played a central role in stellar evolution studies since the initial work of Hayashi, Cameron and Iben in the early 1960s. Hayashi and his contemporaries lacked a theory of the protostar phase, and therefore could not specify with any degree of confidence the initial conditions for PMS contraction. A simple argument indicated that if the interstellar cloud radiated away little energy during its collapse, the starting values of the stellar radius would be some two orders of magnitude greater than the corresponding main-sequence values (Hayashi [20]). The contraction time (Kelvin-Helmholtz time) is so short for these large radii and high surface luminosities that the poorly understood early collapse would be much too brief to affect the subsequent evolution. Adopting the large initial radii, Iben [25] and Ezer & Cameron [15] followed numerically the hydrostatic contraction of stars with masses from less than 1 $M_\odot$ to over 100 $M_\odot$. The evolution of all stars was thought to be qualitatively the same, with quantitative differences scaling in a simple manner with the stellar mass M_*.

1.1 Basic Results

Let us first follow the arguments used by Hayashi and others to derive the initial stellar radii. A protostar of mass M_* and radius R_* forms out of cold, nearly static cloud material whose dimensions are much larger than R_*. Thus, the initial energy (mechanical and thermal) is equal to zero. The star itself is a gravitationally bound entity with a negative total energy. Some of the energy difference is radiated into space during collapse, while most of the rest goes into dissociating and ionizing hydrogen and helium. This internal component, $E_{\rm int}$, is given by

$$E_{\rm int} \equiv \frac{X\,M_*}{m_H}\left[\frac{E^{H_2}_{\rm diss}}{2} + E^{H}_{\rm ion}\right] + \frac{Y\,M_*\,E^{He}_{\rm ion}}{4\,m_H} . \qquad (1)$$

Here, $E^{H_2}_{\rm diss} = 4.48$ eV is the dissociation energy of H_2 molecules, $E^{H}_{\rm ion} = 13.6$ eV is the ionization potential of hydrogen, and $E^{He}_{\rm ion} = 75.0$ eV is the energy needed for full ionization of helium. The hydrogen and helium mass fractions are taken to be X=0.70 and Y=0.28, respectively. Since the protostar's thermal energy is equal to half the gravitational potential energy by the virial theorem, we may write

$$0 = -\frac{1}{2}\frac{G\,M_*^2}{R_*} + E_{\rm int} + L_{\rm rad}\,t\ , \qquad (2)$$

where $L_{\rm rad}$ is the average luminosity escaping over the formation time t. Suppose that we first take the extreme step of ignoring $L_{\rm rad}$ entirely. Then eq. (2) yields the *maximum* radius $R_{\rm max}$ which the protostar could have at any mass M_*. We thus find

$$R_{\rm max} = \frac{G\,M_*^2}{2\,E_{\rm int}} = 61\ R_\odot\left(\frac{M_*}{M_\odot}\right) . \qquad (3)$$

A better treatment of the gravitational energy term changes slightly the numerical value of $R_{\rm max}$. Athough the actual value of the radius of solar-type protostars is still unknown, we know that even the youngest and largest T Tauri stars have radii smaller by an order of magnitude than that predicted by eq. (3).

Having established the initial conditions, the classical theory of PMS evolution then predicts that any star contracts due to heat loss at a rate given by

$$L_{\rm surf} = 4\pi R_*^2 \sigma T_{\rm eff}^4\ , \qquad (4)$$

where $L_{\rm surf}$ is the surface luminosity and $T_{\rm eff}$ the effective temperature of the star. The star is also fully convective because of the large radius. For a one solar mass star with an effective temperature of about 4000 K, $L_{\rm surf}$ ~600 $L_\odot$ which is much higher than the luminosity that can be carried out radiatively

$$L_{\rm rad} = L_0\left(\frac{M_*}{M_\odot}\right)^{11/2}\left(\frac{R_*}{R_\odot}\right)^{-1/2} , \qquad (5)$$

where the precise value of L_0, a luminosity of order 1 $L_\odot$, depends on the star's detailed structure and where it has been assumed that the interior opacity obeys Kramer's Law (Cox & Giuli [10]). Thus, convection spreads inward from the surface (Von Sengbusch [51]).

A fully convective star represents a particularly simple case to model the interior. If the ratio of specific heats is constant, as in the case of a completely ionized gas, the equation of state obeys the relation

$$P = K\rho^{\gamma} = K\rho^{(1+1/n)} \tag{6}$$

where K is a constant and $\gamma = 5/3$ for a monoatomic gas. In such case, the polytropic index $n = 1/(\gamma - 1)$ takes the value 3/2. The gravitational energy Ω of a polytropic star is

$$\Omega = -\frac{3}{5-n}\frac{GM_*^2}{R_*} = -\frac{6}{7}\frac{GM_*^2}{R_*} \,. \tag{7}$$

Thus, the rate at which contraction proceeds given by eq. (7) can also be written as

$$L_{\rm surf} = -\frac{1}{2}\frac{d\Omega}{dt} = -\frac{3GM_*^2}{7R_*^2}\frac{dR_*}{dt} \,, \tag{8}$$

where we have assumed that the loss of radiation occurs at constant mass. Since $L_{\rm surf}$ is positive, $\dot{R}_*$ must be negative, and the star must contract to supply the needed luminosity. The characteristic time scale for the evolution is set by equating eq. (4) to eq. (8),

$$t_{\rm PMS} \equiv -\frac{R_*}{dR_*/dt} = \frac{3GM_*^2/7R_*}{4\pi R_*^2\sigma T_{\rm eff}^4} \approx \frac{3}{7}\frac{GM_*^2}{R_\circ L_\circ} \equiv t_{\rm KH} \,. \tag{9}$$

The last term represents the Kelvin-Helmholtz time scale, and $L_\circ$ corresponds to the luminosity of the star when the star has a fiducial radius $R_\circ$.

The path followed by any star in the H-R diagram is the same. Stars contract *homologously* descending along nearly vertical tracks, commonly called *Hayashi tracks.* The time taken by a star of mass M_* to contract to radius R_* along the Hayashi track is of the order $t_{\rm KH} = 10^7\ M_*^2/R_*^3$ yr. Since the Kelvin-Helmholtz time scale is much longer than the sound crossing time, the star can effectively be considered to be in hydrostatic equilibrium at each instant, despite its slow, or *quasi-static*, contraction.

At a certain point during contraction, the star can no longer remain fully convective. The physical reason is that as the stellar luminosity decreases, the thermal gradient becomes stable and in the central regions a radiative core develops. At this time, the luminosity is no longer determined by the surface boundary conditions, but by the flux in the inner regions. The net result is that the luminosity remains nearly constant as the contraction proceeds. From eq. (4), the effective temperature must increase to compensate for the reduction in radius, while the radiative zone extends towards the surface. Thus, the star moves away from the Hayashi track along a radiative track, first discovered by Henyey, LeLevier & Levee [22] in 1955. Eventually, hydrogen ignites in the center and contraction is halted: the PMS star has reached the *Zero-Age-Main-Sequence* (ZAMS). These results were obtained in the pioneering numerical models of Iben [25] and Ezer & Cameron [15] which established the *standard* set of PMS tracks in the H-R diagram and which can be found in any textbook on stellar evolution.

Before concluding this introductory section, two fundamental properties discovered in the original derivation by Hayashi must be stressed. First, stars of a given mass and radius have a minimum surface temperature, the *Hayashi temperature*; second, the Hayashi temperature is insensitive to the stellar radius, so that the convective portions of the tracks are nearly vertical. Both properties reflect the behavior of the opacity of the gas in the outer layers of the star. The minimum surface temperature is due to the steep decline of opacity with falling temperature below the hydrogen ionization zone near 10^4 K. For too low a surface temperature, the outer layers become optically thin, violating the condition that the photosphere be located at an optical depth of order unity from the surface. The extreme temperature sensitivity of the opacity is also responsible for the vertical nature of the Hayashi tracks. At temperatures below 10^4 K, the opacity varies as T^{α} with α between 7 and 13. Therefore, only a minor change of $T_{\rm eff}$ is needed for a substantial change in the photospheric pressure and entropy (which determine the boundary conditions). For solar-type stars, the effective temperature is about 4000 K. An excellent explanation of the role of the photosphere and entropy structure in determining the basic properties of PMS stars, together with the history of the early developments of PMS theory can be found in the review by Stahler [46].

2 The Impact of Star Formation

The formation of stars occurs within the densest regions of molecular clouds. The detection of infrared point sources embedded in these dense cores shows unequivocally this fundamental property. However, star formation is not simply the result of clouds breaking apart into these dense substructures. The onset of collapse is rather a highly localized occurrence within large complexes, and the character of collapse dictates the structure of the nascent protostar. It is also true that individual collapses can occur over extensive regions of a complex which give birth to large collection of stars in multiple systems, groups, associations and clusters, in order of increasing hierarchy. The observational evidence is that most clouds are not in a state of collapse: support against gravity comes from a balance of forces that persists over long periods of time.

The main support is provided partially from thermal pressure, but also from turbulent motions and the interstellar magnetic field, especially on the largest scales. Typically, the star formation efficiency, defined as the ratio of the mass in stars and the total mass (gas+stars), is of order of few percent in giant molecular clouds. In localized dense clumps, the conversion of gas into stars can be very efficient, reaching values greater than 20-30%, and gravitationally bound clusters of stars can form. The dual character of star formation that gives rise to either isolated single and binary stars or to groups and clusters has come out clearly from a large body of observations of the most spectacular nearby star forming regions.

The formation of a protostar from the quiescent conditions typical of dense cores occurs through the gradual accumulation of the interstellar gas onto an accreting core. This process requires a large decrease in gravitational potential energy. A large fraction of this energy is radiated away in radial or disk accretion shocks that form as a result of the abrupt change in the velocity of the freely-falling gas. If no net energy is absorbed by the circumstellar material, the resulting luminosity is given to a good approximation by

$$L_{\mathrm{acc}} = \frac{GM_*\dot{M}_{\mathrm{acc}}}{R_*} , \tag{10}$$

where M_* and R_* are the instantaneous mass and radius of the protostellar core, and $\dot{M}_{\mathrm{acc}} = dM_*/dt$ the mass accretion rate. Thus, estimates of the luminosity emitted during this phase rely on the knowledge of two fundamental quantities: *the mass accretion rate*, and the *mass-radius relation*. The former is determined by the dynamics of the gravitational collapse, while the relation between the mass and the radius are established by processes occurring in the protostellar interior. The radiation produced at the shock is absorbed, reradiated and thermalized in the optically thick dusty infalling envelope. Most of the observable radiation is emitted at mid- and far-infrared wavelengths. The exact shape of the emergent spectrum depends on the density and temperature distribution of the dust component, which are set by the dynamics of collapse. This section describes how current models of protostellar formation and evolution have constrained both the value of the mass accretion rate, and the structural properties of the central core.

2.1 Protostellar Collapse: the Mass Accretion Rate

An estimate of $\dot{M}_{\mathrm{acc}}$ requires the solution of the dynamical collapse problem. In the idealized case of the collapse of a marginally unstable cloud, such a solution has been found semi-analytically by Shu [44]. In this theory, the rate at which a protostar is built up is

$$\dot{M}_{\mathrm{acc}} = \alpha \frac{a_{\mathrm{T}}^3}{G} . \tag{11}$$

In eq. (11), a_{T} is the effective isothermal sound speed, G the gravitational constant, and α a constant of order unity. If the cloud is supported only by thermal pressure, the expression for the sound speed is simply $a_{\mathrm{T}} = \sqrt{(k_B T/\mu m)}$, where T is the gas kinetic temperature. Otherwise, a_{T} should also include the contribution of magnetic and turbulent pressures. In general,

$$\dot{M}_{\mathrm{acc}} \simeq \frac{(a_{\mathrm{T}}^2 + v_{\mathrm{A}}^2 + v_{\mathrm{turb}}^2)^{3/2}}{G} \tag{12}$$

In dense cores with very little nonthermal support, the gas kinetic temperature varies between 10 and 30 K, corresponding to $a_{\rm T} = 0.1 - 0.3$ km s^{-1}. This yields an average accretion rate of order $\dot{M}_{\rm acc} \sim$ several$\times 10^{-6} - 10^{-5}\, M_{\odot}{\rm yr}^{-1}$. Then, for (sub)solar protostars, the duration of the embedded phase is expected to last typically

$$t_{\rm acc} = M_{*}/\dot{M} \gtrsim 10^{5}\ {\rm yr} \tag{13}$$

The remarkable property of the form of $\dot{M}_{\rm acc}$ in eq. (11) is that it depends only on one parameter, the sound speed, which is controlled by the conditions in the parent cloud, and not by the properties of the central protostar. Equation (11) also implies that the higher the temperature of the cloud, the faster the accretion. This situation is in fact radically different from the more familiar case of the Bondi-type solution of the accretion problem, in which the mass accretion rate does depend on the mass of the star, and is inversely proportional to the velocity (and hence the temperature) of the gas. In Shu's solution, a faster accretion rate implies a higher initial density at the onset of gravitational collapse. The predictions of the *inside-out* collapse model, and the estimate of $\dot{M}_{\rm acc}$, have been extensively used as a diagnostic tool in the interpretation of molecular cloud observations.

2.2 Classical Collapse Models

The onset of the dynamical collapse of a self-gravitating cloud requires the specification of the initial conditions. In the first numerical models (e.g. Larson [26]), these were assumed to correspond to the physical conditions in interstellar clouds at the end of the fragmentation phase, a process by which a massive cloud breaks up into smaller and smaller subunits, clumps and fragments. Considering thermal and gravitational effects alone, the requirement that the clumps be self-gravitating leads to the determination of a critical mass, the well-known Jeans criterion. For an isothermal, non-magnetic, nonrotating sphere composed primarily of molecular hydrogen, neutral helium and heavy metals in solar proportions, the critical wavelength of a perturbation is given by

$$\lambda_J = \left(\frac{\pi\, a_T^2}{G\,\rho}\right)^{1/2} = 0.2\ {\rm pc} \left(\frac{T}{10\ {\rm K}}\right)^{1/2} \left(\frac{n_{H_2}}{10^4\ {\rm cm}^{-3}}\right)^{-1/2} \tag{14}$$

where ρ and n are the mass density and the total number of particles per unit volume. Those perturbations with wavelength exceeding this *Jeans length* have exponentially growing amplitudes. Written in terms of the mass enclosed by λ_J, the Jeans critierion becomes

$$M_J = \frac{a_T^3}{\rho^{1/2}\, G^{3/2}} = 1.6\, {\rm M}_{\odot} \left(\frac{T}{10{\rm K}}\right)^{3/2} \left(\frac{n_{H_2}}{10^4\, {\rm cm}^{-3}}\right)^{-1/2} \tag{15}$$

The numerical factors in equations (14) and (15) show that typical dense cores are close to the edge of gravitational instability. In fact, those with internal stars have already crossed this threshold, and at least some of these ought to show signs of ongoing collapse. Conversely, any structures with measured masses substantially less than M_J or sizes less than λ_J could be stable, given sufficient external pressure, or they could be temporary configurations of positive total energy that will soon cool or disperse. Thus, Larson suggested that protostellar clouds should have initially an almost uniform density, and that the initial density would be close to the critical one.

Starting with these initial conditions, the numerical simulations of protostellar collapse have identified three main stages: an initial *isothermal* contraction, during which the gravitational potential energy is effectively radiated away by the dust and the collapse proceeds at constant temperature; an *adiabatic* phase, corresponding to the transition from low to high optical depth, as a result of the increased density; and, finally, the formation of an *hydrostatic core*, that grows in mass as the collapse proceeds. The basic physical processes that govern each of these phases have been extensively reviewed and need not be repeated here (see Bodenheimer [5]). Suffice to mention only the most general ones, and underline the important fact that they do not depend on the restrictive assumptions under which the numerical calculations were carried out originally (mainly, spherical symmetry). More realistic 2- and 3-dimensional collapse models with less stringent geometry constraints have been developed in the meantime (e.g. Boss [7], Bate [3]), but the main results of the simple case can still be used as useful guidelines.

Gravitational collapse occurs in a highly *non-homologous* way. The central regions of the cloud collapse faster than the outer parts, pressure gradients develop fast, and the nonhomology sets in already during the initial isothermal phase. Independent of the initial conditions, whether uniform or centrally condensed, the matter develops a density distribution of the type $\rho \propto r^{-2}$, where r is the radial distance from the cloud center. In a time slightly longer than the free-fall time, the structure of the cloud resembles that of an asymptotically singular isothermal sphere, modified at the center by the presence of a stable hydrostatic core, which contains only few percent of the total cloud mass. Its instantaneous mass is $M_{\rm p} = \dot{M}_{\rm acc} t$, where t is the time elapsed since collapse. The core is bounded by an accretion shock where the infalling gas is brought to rest. Matter accretes from a freely-falling inner envelope, characterized by a density and velocity profile of the type $\rho \propto r^{-3/2}$ and $v \propto r^{-1/2}$. The outer parts of the clouds are in nearly static equilibrium, with $\rho \propto r^{-2}$.

On the basis of these results, Shu[44] has shown that the singular isothermal sphere,

$$\rho = \frac{a_{\rm T}^2}{2\,\pi\,G\,r^2}\,, \qquad (16)$$

has a self-similar analytic solution that reproduces well the dynamical collapse model. Since a power-law of this type does not have a typical scale, the solution does not define a characteristic mass scale to be associated with the formation of an ordinary star. What the self-similar solution defines is the *rate* at which the central star is built up by accretion. It is exactly this rate that appears in eq. (11). The collapse is said to proceed from *inside-out*, since the inner parts collapse first, and the outer parts later on after being reached by an expansion wave that propagates outward at the sound speed, a_T.

Recently, some concern has been raised about the assumption of the singular isothermal sphere as the initial conditions for the collapse phase. The main criticism is that the density profile has a singularity at the origin, where the density is infinite. Also, in any actual cloud the pressure does not fall to zero, but to some finite value P_ext characterizing the external medium. In such conditions, the critical value of the gravitationally stable mass is given by (Bonnor [6]; Ebert [12])

$$M_\mathrm{BE} = 1.18 \frac{a_\mathrm{T}^4}{P_\mathrm{ext}\, G^{3/2}} = 0.35\,\mathrm{M}_\odot \left(\frac{T}{10\mathrm{K}}\right)^2 \left(\frac{P_\mathrm{ext}}{10^6\, k_B}\right)^{-1/2}, \tag{17}$$

where P_ext is the total pressure, the sum of thermal, turbulent, and magnetic contributions. In a Bonnor-Ebert marginally stable cloud the central density is finite, with a maximum contrast between the central and the ambient values of 14.3. Observations in a variety of dense cores have shown that indeed the density profile flattens out at small radii (André et al. [1], Alves et al. [2]). In addition, the assumption of an isothermal equation of state, while appropriate in dense cores where low-mass star formation occurs, may not be compatible with the conditions observed in the subcondensations of giant molecular clouds that spawn high-mass stars and/or stellar clusters.

As an illustration of the modifications introduced by a different equation of state, we briefly discuss the results obtained in the case of polytropic relations as in eq. (6). Here, however, the index n can also be negative (see McLaughlin & Pudritz [33]; Galli et al. [18]). For a polytropic equation of state with $n < 0$, the sound speed $a_\mathrm{T} \equiv (dP/d\rho)^{1/2} \propto \rho^{1/2n}$ increases with decreasing density, a behavior that appears to accord well with the empirical line width-density relation for molecular clouds (Larson [27]). Since observations indicate that $\Delta v \propto \rho^{-q}$, with q ~0.5 for low-mass cores (Fuller & Myers [16]) and q ~0.2 for high-mass cores (Caselli & Myers [8]), the appropriate value of n would be in the range -1 to -3.

The interesting case of $n = -1$ (the "logatropic" equation of state) in which $P \propto \ln \rho$ has been considered by McLaughlin & Pudritz [34] (see also Lizano & Shu [28]) as the best phenomenological description of the internal structure of clouds and clumps. The predicted density profile is $\rho \propto r^{-1}$ for the static gas, and $\rho \propto r^{-3/2}$ behind the expansion wave where the velocity field is the same as that in free-fall. The basic change is in the mass accretion rate that now is time-dependent with $\dot{M}_\mathrm{acc} \propto t^3$. Thus, the value of the accretion rate is smaller than the isothermal case at early times and larger at late times. The density profiles of the pressure-truncated logatrope and of

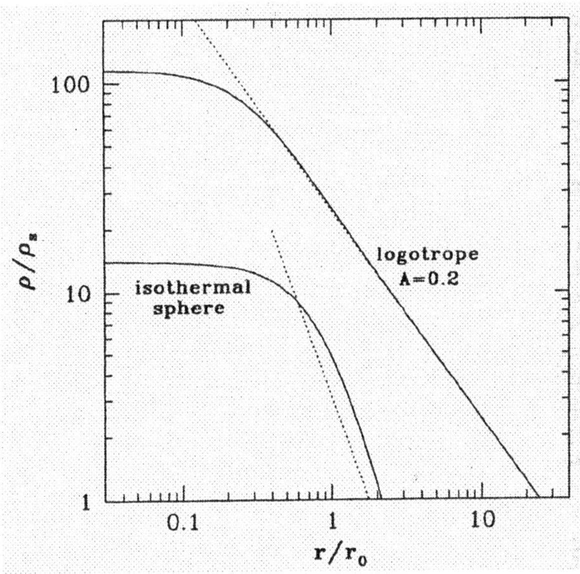

Fig. 1. Density profiles of a pressure truncated logatropic and of the Bonnor-Ebert isothermal sphere. Both configurations are in hydrostatic equilibrium and marginally stable. The dotted lines show the corresponding singular profiles. (From McLaughlin & Pudritz [34])

the Bonnor-Ebert isothermal sphere for a marginally stable cloud are shown in Fig. 1.

Models of protostellar collapse starting from different density profile have also been developed by Henriksen et al. [21] and the result is that the initial phases are characterized by mass accretion rates much faster ($\dot{M}_{\rm acc} \sim 10^{-3}$ $M_\odot$ yr^{-1}) than in the singular isothermal case. However, the high luminosities implied by such accretion rates have not been observed in the youngest, most embedded sources (Class 0 objects ; André et al. [1]). In addition, evidence of cores with density profiles $\rho \propto r^{-2}$ over a significant range of radii is also available (e.g. Evans [14]). In conclusion, the question of the most appropriate initial density profile to describe core collapse is still an open issue.

2.3 More Realistic Models

The assumptions that enter in the derivation of $\dot{M}_{\rm acc}$ represent an oversimplification of the realistic case where forces other than thermal compete in maintaining the stability of molecular clouds. It is clear that molecular clouds cannot be collapsing as a whole on a dynamical time scale, as the Jeans criterion would predict, or the resulting star formation rate in the Galaxy would be far too high. Thus, the concept of the Jeans mass appears of little use in the definition of the appropriate initial conditions, and eqs. (15) and (17) should be replaced by a more realistic one. Among the various mechanisms of cloud support, magnetic fields are the most likely to play the dominant role. A magnetic field of strength B can support a molecular cloud of radius R, provided its mass is less than a magnetic critical mass

$$M_{\rm cr}^{B} = 0.12\frac{\Phi}{\sqrt{G}} \simeq 10^{3}\left(\frac{B}{30\mu G}\right)\left(\frac{R}{2pc}\right)^{2} M_{\odot}\,, \qquad (18)$$

where Φ is the magnetic flux through the cloud (Mouschovias [37]). Depending on the cloud mass, $M_{\rm cl}$, two situations can be envisaged. Supercritical clouds ($M_{\rm cl} \gg M_{\rm cr}$) cannot be supported by magnetic fields alone even if they were perfectly frozen in the gas (no magnetic dissipation), and they would collapse on a magnetic diluted free-fall time scale. Subcritical clouds ($M_{\rm cl} \ll M_{\rm cr}$) are supported by the magnetic field against collapse (even if the external pressure increases) and evolve on the time scale that characterizes the diffusion of the magnetic field. Observational evidence that magnetic fields of the magnitude required by eq. (18) are present in dense cores and in molecular cloud complexes has been obtained by Zeeman splitting measurements (Crutcher [11]).

At the scale of the dense cores, the value of $M_{\rm cr}^{B}$ is of the same order as M_J, namely a few solar masses. The main difference between the two cases is in the time scale of evolution. In dense cores, the gas is lightly ionized (the fractional ionization is about 10^{-7}), and the relevant mechanism of magnetic diffusion is due to *ambipolar diffusion* (Mestel & Spitzer [36]), a process in which the fluid of charged particles can slowly drift with respect to the fluid of neutral particles, the two fluids being coupled by collisions of atomic and molecular species. Studies of the quasi-static evolution of molecular cores have shown that the configuration tends to acquire the density profile of the singular isothermal sphere at the time the central regions become gravitationally unstable and undergo dynamical collapse (Lizano & Shu [28]). Indeed, observations towards dense cores in the Taurus-Auriga region have revealed a high degree of central condensations. On the other hand, the process of ambipolar diffusion is negligible in cloud envelopes because of the relatively high degree of ionization ($x_i \gtrsim 10^{-5}$).

A good approximation for the density of ions, valid in a wide range of molecular cloud densities, is $\rho_i = C\rho_n^{1/2}$, where ρ_i and ρ_n are the ion and neutral densities, respectively, and $C = 3\times10^{-16}$ cm$^{-3/2}$ g$^{1/2}$ is a constant. Under these conditions, ambipolar diffusion sets in gradually, and its characteristic time scale in a molecular cloud core of size $R_{\rm cl}$ and density ρ_n is

$$t_{\rm AD} \sim \frac{R_{\rm cl}}{v_d} \sim \frac{4\pi\gamma C\rho_n R_{\rm cl}^2}{B^2} = 13\left(\frac{n_{H_2}}{10^4\,{\rm cm}^{-3}}\right)^{3/2}\left(\frac{R_{\rm cl}}{0.2\,{\rm pc}}\right)^{2}\left(\frac{B}{30\,\mu{\rm G}}\right)^{-2}\,{\rm Myr}\,, \qquad (19)$$

where v_d is the ion-neutrals drift velocity and $\gamma = 3.5\times10^{13}\,\mathrm{cm^3\,g^{-1}\,s^{-1}}$ is the ion-neutral drag coefficient. For typical molecular cloud cores, the diffusion time is long compared to the dynamical free-fall time, by a factor of ~10 (e.g. McKee et al. [32]).

Collapse calculations starting from these initial conditions have been presented by several authors who considered in detail the effects of magnetic field and plasma drift (e.g., Galli & Shu [17]; Ciolek & Mouschovias [9]). The results indicate that the collapse still takes place in an inside-out fashion, but now propagating outward as a fast magnetosonic wave, traveling faster in the direction parallel to the initial magnetic field. Interestingly, the value of $\dot{M}_{\rm acc}$ obtained in the magnetic case is *not* significantly altered by the presence of a magnetic field. The reason is that the gas is slowed down, because of the Lorentz force opposing gravity, but the information travels outward faster, because of the increased characteristic speed of the system. The two effects cancel out, and the numerical value of $\dot{M}_{\rm acc}$ remains unchanged.

However, the dynamics of the gas is significantly affected by the magnetic field. The strong pinching Lorentz force deflects the infalling gas toward the equatorial plane, with the formation of an inner disequilibrium structure. The disk instantaneous radius is given by

$$\begin{aligned} r_B &\simeq 0.12 \left(\frac{G^2\,B^4}{a_{\rm T}}\right)^{1/3} t^{7/3} \\ &\simeq 600 \left(\frac{B_0}{30\,\mu\mathrm{G}}\right)^{4/3} \left(\frac{a_{\rm T}}{0.3\,\mathrm{km\,s^{-1}}}\right)^{-1/3} \left(\frac{t}{10^5\mathrm{yr}}\right)^{7/3} \; AU. \end{aligned} \tag{20}$$

Disk-like structures over scales of several hundred AUs have been mapped in molecular line emission around many YSOs (Beckwith [4]). The scale of the magnetic pseudo-disk is much larger than that of the centrifugal disk, that forms as a result of the effects of the initial small, but not negligible rotation of the molecular cloud. The value of the centrifugal radius is (Terebey et al. [49])

$$\begin{aligned} r_{\rm cen} &= 0.06\,\left(a_{\rm T}\,\Omega^2\right)\,t^3 \\ &\simeq 7 \left(\frac{a_{\rm T}}{0.3\,\mathrm{km\,s^{-1}}}\right) \left(\frac{\Omega}{4\times10^{-14}\,\mathrm{s^{-1}}}\right)^2 \left(\frac{t}{10^5\,\mathrm{yr}}\right)^3 \; AU\,, \end{aligned} \tag{21}$$

where Ω is the initial rotation rate of the core. Inside $r_{\rm cen}$, infalling matter encounters a centrifugal barrier in the equatorial plane, and accumulates in a disk. The mass of the protostar at this time is thus $M_{\rm proto} = \dot{M}t$, which gives

$$\begin{aligned} M_{\rm proto} &= \left(\frac{16\,R_*\,a_T^8}{G^3\,\Omega^2}\right)^{1/3} \\ &= 0.2\,M_\odot \left(\frac{R_*}{3\,R_\odot}\right)^{1/3} \left(\frac{\Omega}{10^{-14}\,s^{-1}}\right)^{-2/3} \left(\frac{a_T}{0.3\,km\,s^{-1}}\right) \end{aligned} \tag{22}$$

The various elements that characterize the collapse of magnetized, rotating molecular cores over a range of different scales are schematically shown in Fig. 2.

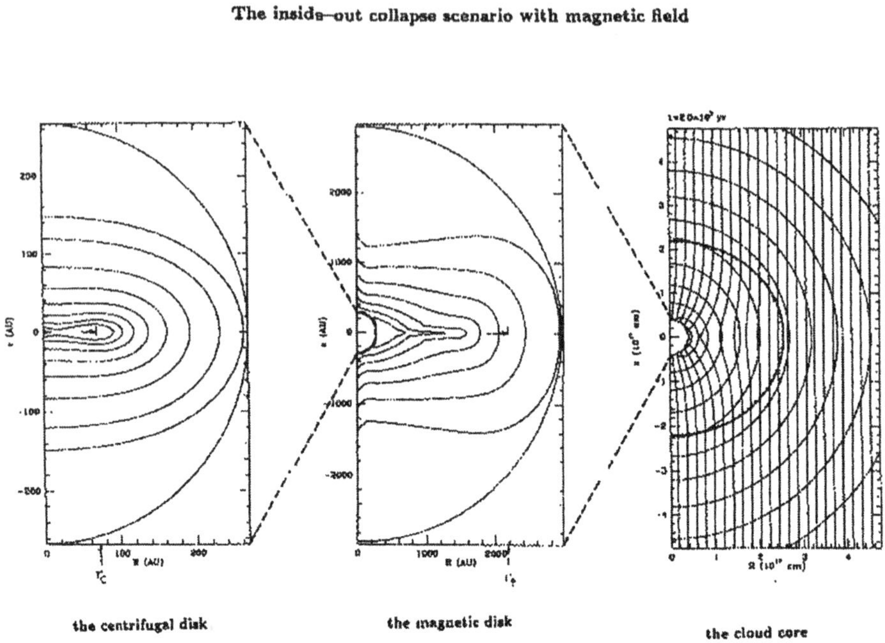

Fig. 2. Global view of the collapse of a magnetized, rotating dense core. The right panel shows the isodensity contours and magnetic field lines over a scale of about 0.1 pc. The dotted line represents the infall magnetosonic wave that propagates faster in the direction perpendicular to the field. In the middle panel, on scales of $\sim 10^3$ AU, the dynamics is dominated by magnetic and gravitational forces that shape the density distribution in the equatorial plane. The left panel shows the isodensity contours in the innermost $\sim 10^2$ AU. The positions of r_B and r_C (see text) are indicated in the middle and left panels. (Adapted from Galli & Shu [17])

3 Protostellar Evolution

Having established the magnitude of the mass accretion rate, we now discuss the properties of protostars, as they grow from the small hydrostatic core formed at the end of the adiabatic phase to a stellar sized object during the main accretion phase. The purpose here is to show how a *mass-radius* is established , that can then be used in eq. (10) to estimate the protostellar luminosity.

In the inside-out collapse scenario, the structure and evolution of the protostar are solely determined by the specification of $\dot{M}_{\rm acc}$. The evolution of the infalling envelope and the hydrostatic core are effectively decoupled, owing to the existence of a region of very low opacity (*the opacity gap*) in the outer parts of the envelope. The photons produced at the accretion shock can escape freely, instead of being pushed back into the core by the infalling matter. This results in a great simplification of the problem, that otherwise would require the joint solution of the full hydrodynamical equations for the dynamical collapse and a detailed radiation transfer. Despite great efforts, fully self-consistent models do not yet exist (see Masunaga et al. [30]; Bate [3]).

The quasi-hydrostatic evolution of the protostar can be treated as a problem in stellar structure, with the modifications introduced by a time varying mass and nonstandard surface boundary conditions provided by an accretion shock (Stahler et al. [48]). These models have shown the key role of the nuclear burning of interstellar deuterium in determining the properties of the accreting core. The protostellar evolution in a mass interval that covers low- and intermediate-mass stars, $0.1 \lesssim M_*/M_\odot \lesssim 10$, is marked by the occurrence of four distinct phases of deuterium burning, independently of the assumptions of the accretion process (Stahler [47]; Palla & Stahler [39]). These phases are sketched in Fig. 3 and will be described in detail below.

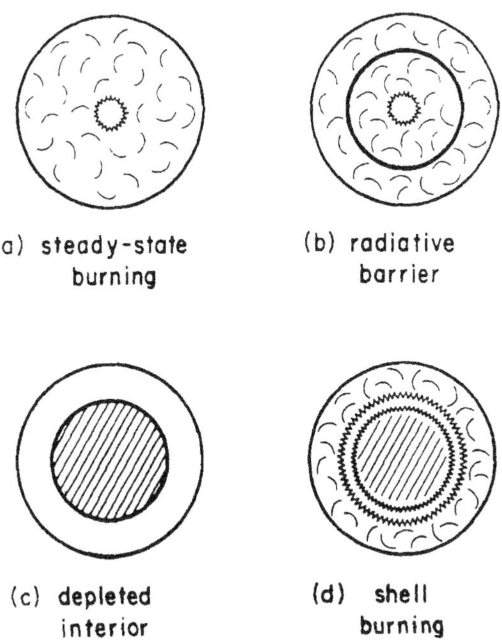

Fig. 3. A schematic view of the main phases of deuterium burning in protostars (from Palla & Stahler [38]).

3.1 Central D-Burning

For low-mass protostars ($M_* \lesssim 1\ M_\odot$), deuterium burning occurs near the center for temperatures exceeding 10^6 K. The energy input is sufficient to turn and maintain the star convectively unstable. As more mass is gathered, the newly accreted deuterium is instantly transported by convective eddies to the center, and a situation of steady-state burning is created. In this condition, the deuterium generated luminosity, L_D is given by

$$L_D \simeq \dot{M}_{\rm acc}\,\delta = 12 L_\odot \left(\frac{\dot{M}_{\rm acc}}{10^{-5}\ M_\odot\ {\rm yr}^{-1}} \right) . \qquad (23)$$

In eq. (23) the nuclear energy per unit mass available in interstellar matter is given by $\delta \equiv [D/H] X Q_H / m_H$, where $[D/H] = 2 \times 10^{-5}$ is the deuterium fractional abundance by number, $X = 0.7$, and $Q_H = 5.5$ MeV is the energy available per reaction. The temperature sensitivity of the energy generation rate

$$\epsilon_D = 4.2 \times 10^7\,[D/H] \left(\frac{\rho}{1\,g\,cm^{-3}} \right) \left(\frac{T}{10^6\,K} \right)^n erg\,g^{-1}\,s^{-1} , \qquad (24)$$

with $n = 11.8$, is such that deuterium acts as an effective thermostat, preventing the central temperature from rising above $T_c \sim 10^6$ K as the star gains mass (Stahler [47]). Since the amount of energy available in deuterium, δ, is comparable to the gravitational binding energy of the star, GM_*/R_*, the thermostatic effect results in a core radius that increases almost linearly with mass during the phase of active burning. This trend is illustrated in Fig. 4 for $0.3 \lesssim M_*/M_\odot \lesssim 1$, assuming a constant accretion rate of 1×10^{-5} $M_\odot$ yr^{-1}. The other curves each have a different value for R_* at the initial mass, here taken to be 0.1 $M_\odot$. It is evident that, by the time M_* has doubled, the curves are nearly identical. The open circle marks the mass when the interior becomes fully convective.

To gauge the evolutionary influence of different mass accretion rates on the resulting mass-radius relation, Fig. 5 shows the results for lower values of $\dot{M}$. In a protostar forming at low $\dot{M}$, the radius at a given mass is relatively small, resulting in a higher central temperature, earlier deuterium ignition and earlier depletion of nuclear material. Thus, the core slows its growth at a small mass. The onset of full convection occurs at smaller masses as $\dot{M}$ is lowered. The mass-radius relation also depends on the deuterium interstellar concentration. The lower panel of Fig. 5 illustrates this point in the extreme case where $[D/H]$ is set equal to zero and three different values of $\dot{M}$ are considered. It is seen that the radius increases smoothly and the curves are well separated by their $\dot{M}$ values. Of course, as soon as a modest abundance of deuterium becomes available, the modest rise in radius will be replaced by a more pronounced increase, due to the thermostatic nature of deuterium burning.

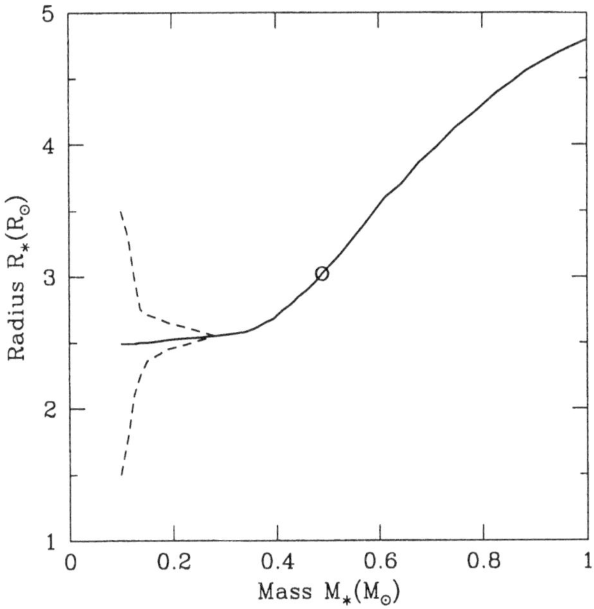

Fig. 4. Mass radius relation for a protostar accreting at $\dot{M} = 10^{-5}\,M_\odot\,yr^{-1}$. As shown by the dashed curves, different initial radii quickly converge to the same mass-radius relation. The circle marks the onset of full convection. (From Stahler [47])

3.2 Radiative Barrier

The evolution of protostars more massive than solar is marked by the return to radiative stability and the onset of gravitational contraction. These events occur at about the same time, as can be seen by the following argument. Convection ceases to be effective when the equality

$$L_D = L_{\rm rad}(M_*, R_*) , \tag{25}$$

where $L_{\rm rad}$ is the luminosity that a star of given mass and radius can carry radiatively, is met. For low-mass stars, $L_{\rm rad}$ is very small, but it quickly rises, due to the strong dependence on stellar mass, $L_{\rm rad} \sim M_*^{5.5}\,R_*^{-1/2}$. The requirement on the equality of the two luminosities provides a condition on the star's gravitational contraction time, t_{KH}, at the time of radiative equilibrium, i.e.

$$t_{\rm KH} \equiv \frac{GM_*^2}{R_* L_{\rm rad}} \approx \frac{GM_*^2}{R_* L_{\rm D}} \approx \frac{GM_*^2}{R_* \dot{M} \delta} \approx \frac{M_*}{\dot{M}} \equiv t_{\rm acc}. \tag{26}$$

Equation (26) establishes that the star is losing heat by radiation at the same time as it accretes new matter, and therefore it undergoes a phase of rapid

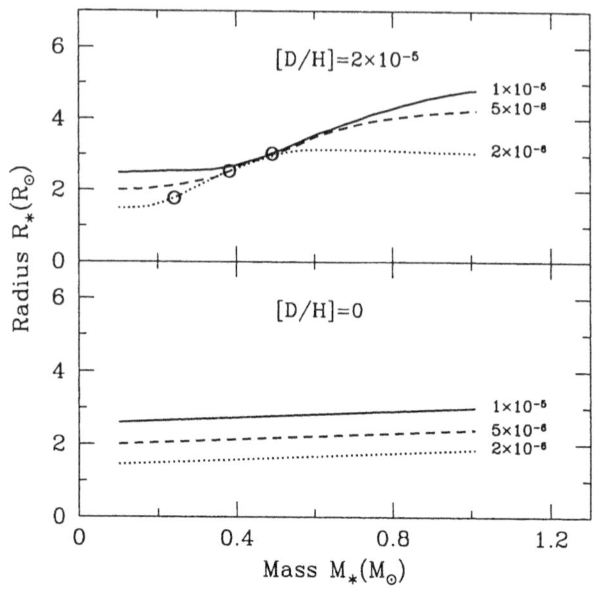

Fig. 5. Dependence of the mass-radius relation on the mass accretion rate and on the initial deuterium concentration. In either case, the three curves are for different values of $\dot{M}$, as indicated. In the upper panel, the open circles mark the onset of full convection. In the lower panel, no deuterium is present in the gas. of $[D/H]$. (Adapted from Stahler [47])

gravitational contraction. Note that for low-mass stars, the accretion time is always much shorter than $t_{\rm KH}$. Due to the sensitivity of $L_{\rm rad}$ on M_*, the equality of the two time scales is confined to a rather narrow range of masses, between 2 and 3 $M_\odot$. If nothing else happened in the interior, the rapid contraction would lead to the conditions appropriate for hydrogen burning in the center, and the star would then join the main-sequence while still accreting. As a consequence, there should be no stars of mass greater than $2-3\ M_\odot$ in the pre-main-sequence phase. This result is in clear contradiction with the observational evidence of intermediate-mass stars, the so-called Herbig Ae/Be stars, whose location in the H-R diagram is well above the ZAMS. The reason why this expected behavior does not occur is subtle, and is again related to the effects of deuterium burning. Physically, the transition to radiative equilibrium does not occur throughout the star all at once, but begins in a localized region (the radiative barrier). Interior to the barrier, matter quickly burns its nuclear fuel that can no longer be replenished by newly accreted deuterium, and an inert core develops (see the second and third panel of Fig. 3).

3.3 The Onset of D-Shell Burning and Gravitational Contraction

Matter exterior to the barrier still contains substantial deuterium. When the thermodynamical conditions are appropriate, deuterium once more ignites, this time in a convective shell located at the barrier, as indicated in the last panel of Fig. 3. This shell slowly expands toward the surface, due to the overall tendency of the star to become more and more radiatively stable. The release of nuclear energy in the lower density subsurface regions results in a dramatic swelling of the star, that doubles its radius. This is shown in the steep, almost vertical rise of the curve in Fig. 6. The expansion of the radius counteracts the gravitational pull, and retards the beginning of the phase of gravitational contraction. As a consequence, the fusion of ordinary hydrogen, and thus the arrival on the main sequence is postponed to higher masses, $\simeq 8\ M_\odot$ in the specific case considered here, thus solving the inconsistency with the observations.

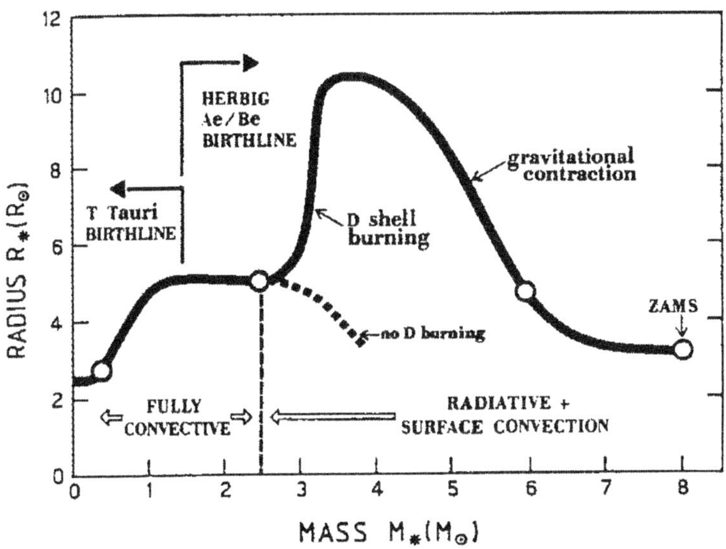

Fig. 6. The evolution of the radius vs. mass in accreting protostars. The open circles represent, from left to right, the ignition of central deuterium, the start of D-shell burning, the ignition of central hydrogen via the CN cycle, and the final arrival on the ZAMS at a mass $M_* \sim 8\ M_\odot$. To gauge the importance of D-shell burning, the dashed curve shows the path of the radius obtained ignoring it (adapted from Palla & Stahler [38]).

3.4 The Stellar Birthline

Under the assumption that protostars end their accretion phase in a time much shorter than the Kelvin-Helmholtz time, the mass-radius relation shown in Fig. 6 can be used to construct a theoretical birthline for stars of low- and intermediate-mass in the H-R diagram. The birthline is the locus along which young stars first appear as optically visible stars (Stahler [45]). Accordingly, there should be no observed PMS stars whose position in the H-R diagram is *above* the birthline, since the largest radius a PMS star of a given mass can achieve is the protostellar radius at that mass.

The resulting birthline is shown in Fig. 7, together with the location of T Tauri stars and Herbig Ae/Be stars. It is evident that the upper envelope of the stellar distribution is generally well matched by the theoretical birthline. The agreement is also suggestive of the fact that the specialized assumptions of the calculations (spherical symmetry, constant mass accretion rate) are not essential to the outcome of the mass-radius relation (see also Hartmann et al. [19]).

Corroboration of these evolutionary calculations with the data on Herbig Ae and Be stars has already begun. For example, stellar properties for all candidate Herbig stars (44 objects) measured by Hipparcos have been derived (van den Ancker et al. [50]). The distribution in the H-R diagram of 16 stars with accurate trigonometric parallaxes (better than 3σ) is shown in Fig. 8. The typical error in log T_{eff} is about 0.05 (or one subclass in spectral type), whereas the error in luminosity is dominated by the error in the distances. With the exception of one star (HD 34282) below the ZAMS, all the others fall below the birthline for 10^{-5} $M_{\odot}$ yr^{-1}. The brightest star is HD 200775 which is probably in a more advanced evolutionary phase than the others which are still contracting toward the main sequence.

An interesting prediction of the birthline is that the *ages* of intermediate mass stars should be much less than previously thought (Palla & Stahler [38]). The predicted endpoint to the birthline at 8 $M_{\odot}$ also appears supported by the observational data. For example, the detailed study of the stellar population of the Trapezium cluster by Hillenbrand [23] clearly shows that stars more massive than $\sim$6 $M_{\odot}$ are grouped tightly about the ZAMS, while less massive ones diverge from it. A very similar pattern is evident in the clusters NGC 6611 (Hillenbrand [24]) and NGC 3603 (Eisenhauer et al. [13]), in a number of Galactic OB associations (Massey et al. [29]), and in the R136 cluster of 30 Doradus (Massey & Hunter [30]).

3.5 Effects of the Accretion Flow

Variations in the geometry of the accretion flow mainly reflect the influence of rotation and magnetic fields in the parent cloud. Angular momentum conservation imposes that the infalling material does not strike the stellar surface directly, but first forms a circumstellar disk from which it spirals into the star

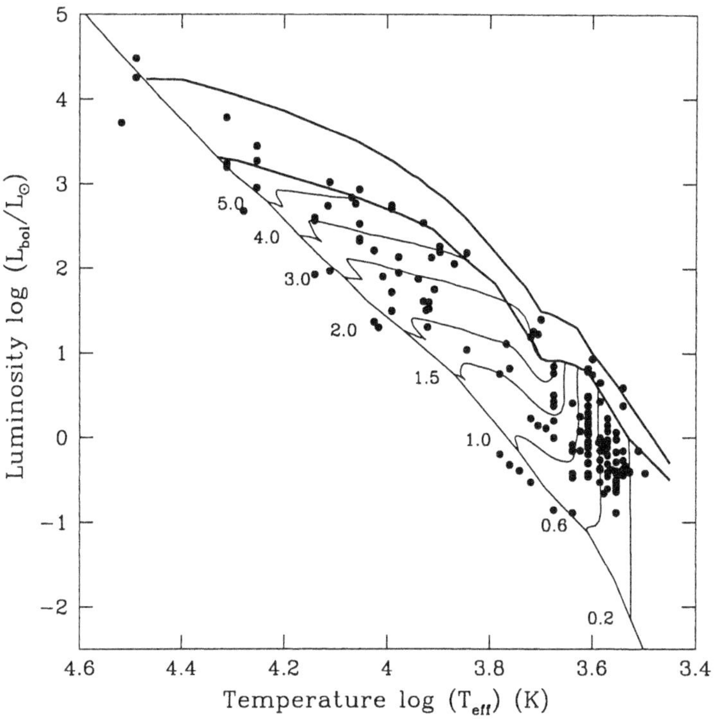

Fig. 7. Location of the stellar birthline in the H-R diagram for accretion rates of 10^{-5} (lower heavy curve) and 10^{-4} (upper heavy curve) $M_{\odot}\mathrm{yr}^{-1}$. Symbols give the positions of young PMS stars. Labeled curves are PMS evolutionary tracks for the indicated stellar mass. (From Palla & Stahler [41])

(e.g. Mercer-Smith et al. [35]). The largest uncertainty in accounting for this effect in numerical models comes from the lack of detailed knowledge of the physical mechanism responsible for the accretion through the disk, and an *ad hoc* prescription must be used. Qualitatively, the global effect of accretion through a disk is to yield matter landing onto the star with a lower specific entropy than that which hits the surface directly, because of the heat loss through radiation from the disk faces. In actual calculations, where this effect has been mimicked by altering the boundary conditions for the core, the resulting protostellar radius results smaller than in the shock case. However, as shown in Fig. 9, the sequence of events described above remains unaltered with the protostar undergoing the same dramatic swelling at the time of deuterium shell-burning (Palla & Stahler [40]).

As for a variation of the mass accretion rate, there are observational indications that stars of increasing mass may be formed from molecular cloud cores with large nonthermal internal motions. These motions would provide support to the cloud against collapse, and would tend to increase $\dot{M}_{\rm acc}$ well

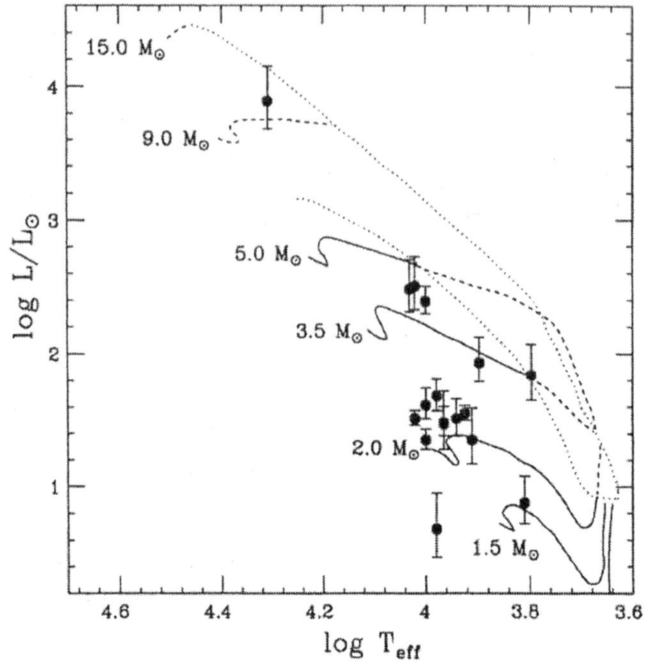

Fig. 8. H-R diagram of Herbig Ae/Be stars with parallaxes measured by Hipparcos. Also shown are the PMS evolutionary tracks and birthlines for 10^{-5} and 10^{-4} $M_\odot$ yr^{-1} from [41]. (From van den Ancker et al. [50])

above the thermal value given by eq. (2). On the other hand, if the fiducial accretion rate was much *lower* than the thermal one, the resulting intermediate-mass stars would run through the entire PMS evolution during protostellar accretion, a result in clear conflict with the observations. The effect of increasing $\dot{M}$ is opposite to that of assuming disk accretion; namely, the entropy level of the protostar becomes higher, since each gas element landing on the star has less time to radiate away its heat. The dependence of the mass-radius relation on the accretion rate is illustrated in Fig. 10. The various evolutionary phases are once again recovered: the shape of the mass-radius remains unaltered, even though each event is shifted to higher masses.

The fact that $R_*(M_*)$ reaches a maximum and then declines is a common property of accreting protostars and originates from the combined effects of deuterium burning in a shell and the eruption of the luminosity wave from the interior. The behavior of the radius can be understood in the following terms: for small masses, nuclear heating causes $M_*(R_*)$ to increase linearly (M_* =0.5–1 $M_\odot$), while at larger masses (M_* =1–2 $M_\odot$) the self-gravity of the star takes over, and the radius declines as $M_*^{-1/3}$. A useful expression that relates the peak radius at maximum expansion to the pre-expansion

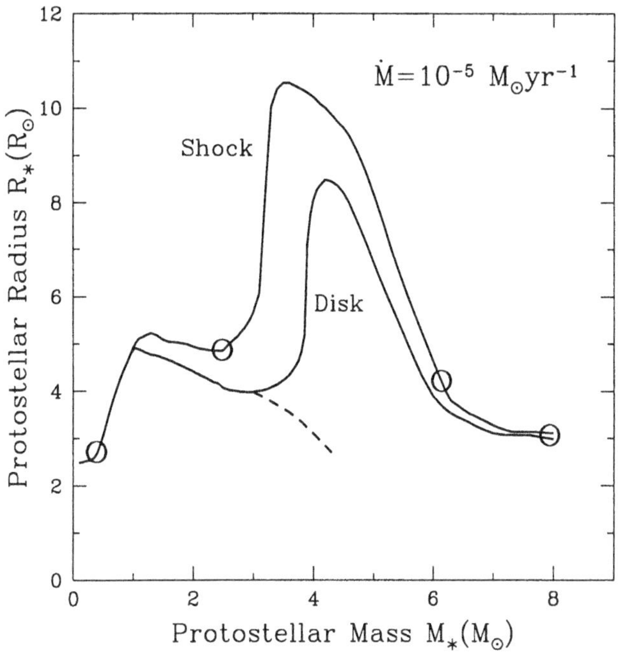

Fig. 9. The evolution of the radius vs. mass in accreting protostars. The two curves are for the case of radial infall and disk accretion at the fiducial rate 10^{-5} M$_\odot$ yr $^{-1}$. The open circles mark the evolutionary phases as in Fig. 4 The evolution without deuterium burning is displayed by the dashed curve. (Adapted from Palla & Stahler [40])

radius is given by

$$\frac{R_{\rm max}}{R_{\rm pre}} \approx \left(\frac{L_{\rm max}}{\dot{M}\,\delta}\right)^{1/2}, \tag{27}$$

where $L_{\rm max}$ is the luminosity at maximum radius. In all cases shown in the figure, $L_{\rm max}/\dot{M}\delta \sim 5$, so that the radius swells by a factor of ~2.

The resulting birthline computed with an accretion rate of 10^{-4} M$_\odot$ yr $^{-1}$ is shown in Fig. 7 by the uppermost curve. Clearly, the new curve does not greatly improve the agreement with the observational data. These considerations lead to the conclusion that an accretion rate $\dot{M} \sim 10^{-5} M_\odot {\rm yr}^{-1}$ can represent the typical value for the formation of stars in a mass range from few tenths of a solar mass to about 10 $M_\odot$.

References

1. André, P., Ward-Thompson, D., Barsony, M. 2000, in Protostars and Planets IV, eds. V. Mannings, A.P. Boss & S.S. Russell (Tucson: University of Arizona

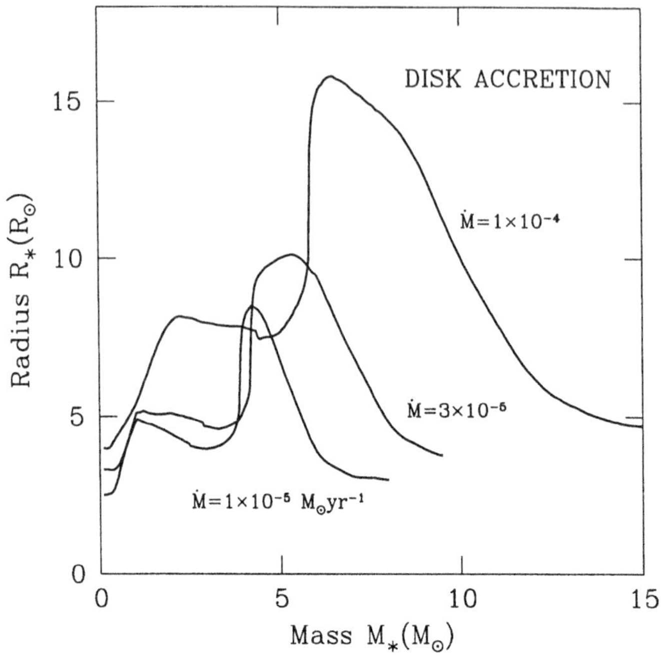

Fig. 10. The dependence of mass–radius relation on the accretion rate. The values of $\dot{M}_{\rm acc}$ considered are for the case of disk accretion. (Adapted from Palla & Stahler [40])

Press), p. 59
2. Alves, J.F., Lada, C.J., Lada, E.A. 2001, Nature, 409, 159
3. Bate, M.R. 1998, ApJ, 508, L95
4. Beckwith, S.V.W. 1999, in The Origin of Stars and Planetary Systems, eds. C.J. Lada & N.D. Kylafis (Dordrecht: Kluwer Acad. Press), p.577
5. Bodenheimer, P. 1992, in Star Formation in Stellar Systems, eds. G. Tenorio-Tagle, M. Prieto & F. Sánchez (Cambridge: Cambridge Univ. Press), p.1
6. Bonnor, W.B. 1956, MNRAS, 116, 351
7. Boss, A.P. 1999, ApJ, 520, 744
8. Caselli, P., Myers, P.C. 1995, ApJ, 446, 665
9. Ciolek, G.E., Mouschovias, T.Ch. 1996, ApJ, 468, 749
10. Cox, J.P., Giuli, R.T. 1968, Principles of Stellar Structure, (New York: Gordon and Breach)
11. Crutcher, R.M. 1999, ApJ, 520, 706
12. Ebert, R. 1955, Z. Astr. 37, 222
13. Eisenhauer, F. Quirrenbach, A., Zinnecker, H., Genzel, R. 1998, ApJ, 498, 278
14. Evans, N.J.,II 1999, ARAA, 37, 311
15. Ezer, D., Cameron, A.G.W. 1965, Canadian J. Phys., 43, 1497
16. Fuller, G.A., Myers, P.C. 1992, ApJ, 384, 523
17. Galli, D., Shu, F.H. 1993, ApJ, 417, 220
18. Galli, D., Lizano, S., Li, Z.-Y., Adams, F.C., Shu, F.H. 1999, ApJ, 521, 630

19. Hartmann, L., Cassen, P., Kenyon, S.J. 1997, ApJ, 475, 770
20. Hayashi, C. 1966, ARAA, 4, 171
21. Henriksen, R., André, P., Bontemps, S. 1997, A&A, 323, 549
22. Henyey, L.G., LeLevier, R., Levee, R.D. 1955, PASP, 67, 154
23. Hillenbrand L.A. 1997, AJ 113, 1733
24. Hillenbrand L.A., Massey P., Strom S.E., Merrill K.M. 1993, AJ, 106, 1906
25. Iben, I. 1965, ApJ, 141, 993
26. Larson, R.B. 1969, MNRAS, 145, 121
27. Larson, R.B. 1981, MNRAS, 194, 809
28. Lizano, S., Shu, F.H. 1989, ApJ, 342, 834
29. Massey P., Johnson K.E., DeGioia-Eastwood K. 1995, ApJ, 454, 151
30. Massey P., Hunter D.A. 1998, ApJ, 493, 180
31. Masunaga, H., Miyama, S.M., Inutsuka, S. 1998, ApJ, 495, 346
32. McKee, C.F., Zweibel, E.G., Goodman, A.A., Heiles, C. 1993, in Protostars and Planets III, eds. E.H. Levy & J.I. Lunine (Tucson: University of Arizona Press), p.327
33. McLaughlin, D.E., Pudritz, R.E. 1996, ApJ 469, 194
34. McLaughlin, D.E., Pudritz, R.E. 1997, ApJ, 476, 750
35. Mercer-Smith, J.A., Cameron, A.G.W., Epstein, R.I.: 1984, ApJ, 279, 363
36. Mestel, L., Spitzer, L. 1956, MNRAS, 116, 505
37. Mouschovias, T.Ch., Ciolek, G.E. 1999, in The Origin of Stars and Planetary Systems, eds. C.J. Lada & N.D. Kylafis (Dordrecht: Kluwer Acad. Press), p. 305
38. Palla, F., Stahler, S.W. 1990, ApJ, 360, L47
39. Palla, F., Stahler, S.W. 1991, ApJ, 375, 288
40. Palla, F., Stahler, S.W. 1992, ApJ, 392, 667
41. Palla, F., Stahler, S.W. 1993, ApJ, 418, 414
42. Palla, F., Stahler, S.W. 1999, ApJ, 525, 772
43. Palla, F., Stahler S. W. 2000, ApJ, 540, 255
44. Shu, F.H. 1977, ApJ, 214, 488
45. Stahler, S.W. 1983, ApJ, 274, 822
46. Stahler, S.W. 1988a, PASP, 100, 1474
47. Stahler, S.W. 1988b, ApJ, 332, 804
48. Stahler, S.W., Shu, F.H., Taam, R.E. 1980, ApJ, 302, 590
49. Terebey, S., Shu, F.H., Cassen P. 1984, ApJ, 286, 529
50. van den Ancker M., de Winter, D., Tjin A Djie, H.R.E. 1998, A&A, 330, 145
51. Von Sengbusch, K. 1968, Zs.f.Ap., 69, 79

II. Pre-Main-Sequence Evolution

4 Evolutionary Models

Once the main phase of accretion is completed, the stellar core emerges as an optically visible star along the birthline. The physical process by which infall stops is still not known, although stellar winds and bipolar outflows must play a fundamental role. The circumstellar matter surrounding these young stars, partly distributed in a disk and the rest in an extended envelope, still emits copiously at infrared wavelengths: the emergent spectral energy distribution departs substantially from that of a normal stellar photosphere, showing excess emission extending from the near- to the far-infrared. The origin of this excess has been commonly attributed to the presence of accretion disks. Depending on the mass of the star, young visible stars are classified as T Tauri (low-mass, typically $\sim$1 $M_\odot$), or Herbig Ae/Be stars (intermediate-mass, $\lesssim$10 $M_\odot$). Their early evolution can be followed in detail, once the initial conditions for the gravitational contraction phase toward the Main Sequence are specified. The birthline joins the ZAMS at $M_* \sim$8 $M_\odot$. Thus, for stars with masses above this value there is no optical PMS phase: the rapid gravitational contraction in these objects has led to hydrogen ignition already during the accretion phase. Massive stars are born *on* the ZAMS.

4.1 Protostellar Initial Conditions

The results of protostellar evolution demonstrate that the classical assumptions described in Sect. 1 are no longer appropriate. First, protostellar cores never reach large radii: the radius remains typically a factor of ten smaller than that predicted by eq. (3). It is thus reasonable to assume that the protostellar radius represents the *maximum* value that a star of a given mass can achieve prior to contraction. Second, the internal structure departs significantly, at least for intermediate-mass objects, from the assumption of thermal convection. In low-mass stars, convection is due to central nuclear burning. Protostars more massive than about 2 $M_\odot$ are radiatively stable in the inner regions, and possess a thick, subsurface mantle of deuterium. This deuterium must ignite in a shell to fuse to helium during the approach to the main sequence. The disparate state of the two regions implies that the stars are *thermally unrelaxed* at the beginning of the PMS evolution. Thus, in contrast to the situation described by the classical theory, these stars must undergo *non-homologous* quasi-static contraction (Palla & Stahler [41]).

Another important aspect is that for a PMS of fixed mass, the amount of deuterium available is that which a protostar of the same mass did acquire in the course of its accretion history. The variation of the fractional deuterium concentration, f_D, for a protostar accreting at different rates is shown in Fig. 11. Here, f_D is calculated relative to the interstellar concentration. We see that for the lowest values of the accretion rate, f_D, declines quite rapidly for

$M_* \lesssim 1$ M$_\odot$, even thoguh the star is fully convective, and therefore accreting deuterium as fast as it is being consumed. Actually, consumption slightly exceeds accretion, and L_D differs correspondingly from the value in eq. (23). The evolution of f_D in a fully convective protostar is given by

$$M_* \frac{df_D}{dM_*} = 1 - f_D - \frac{L_D}{\dot{M}_{acc}\delta} , \tag{28}$$

from which it follows that

$$L_D = \dot{M}_{acc}\delta \left[1 - \frac{d(f_D M_*)}{dM_*}\right] . \tag{29}$$

Since the product $f_D M_*$ is proportional to the total deuterium content, equation (29) shows that any net loss of deuterium implies that L_D exceeds $\dot{M}_{acc}\delta$. Such loss is inevitable, because of the slowly rising interior temperature of the protostar. The fall of f_D shown in Fig. 11 continues until the radiative barrier appears, and thereafter deuterium exists only in a small, subsurface mantle. For fast accretion rates, the drop of deuterium in the protostellar phase is postponed at masses $M_* \sim 1.5$ M$_\odot$.

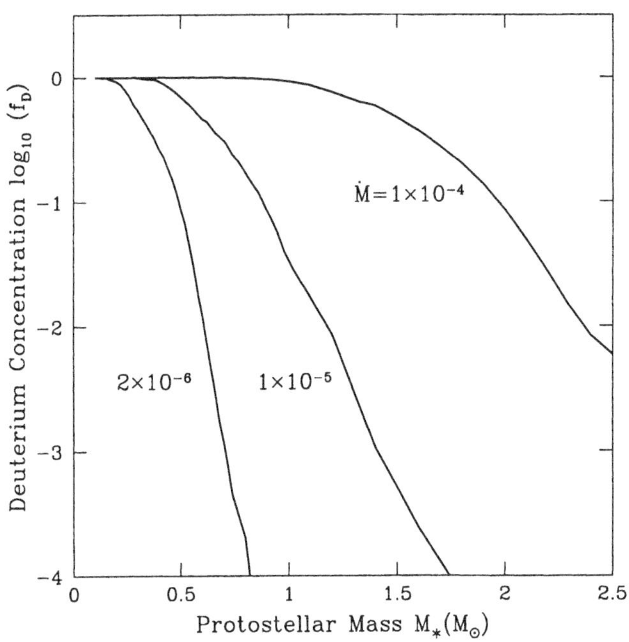

Fig. 11. Evolution of the fractional deuterium concentration, f_D, in protostars accreting at different values of the accretion rate.

The calculation of the buildup of a protostar to a mass M_* automatically yields the initial PMS structures for all stars of lower mass, provided only that the main accretion process ends relatively quickly (compared to the Kelvin-Helmholtz time scale). Since the contraction time is of order 10^7 yr for a solar-type star, while the typical clearing time due to a wind/outflow is 10^5 yr, this assumption is well satisfied. The salient features of the initial phases of PMS evolution can be discerned by comparing several characteristic luminosities as functions of stellar mass, as depicted in Fig. 12. The figure displays the surface luminosity $L_{\rm surf}$ at the start of the PMS phase. Also, Fig. 12 shows $L_{\rm rad}$, the luminosity transported by radiation within the star, and the protostellar accretion luminosity $L_{\rm acc}$.

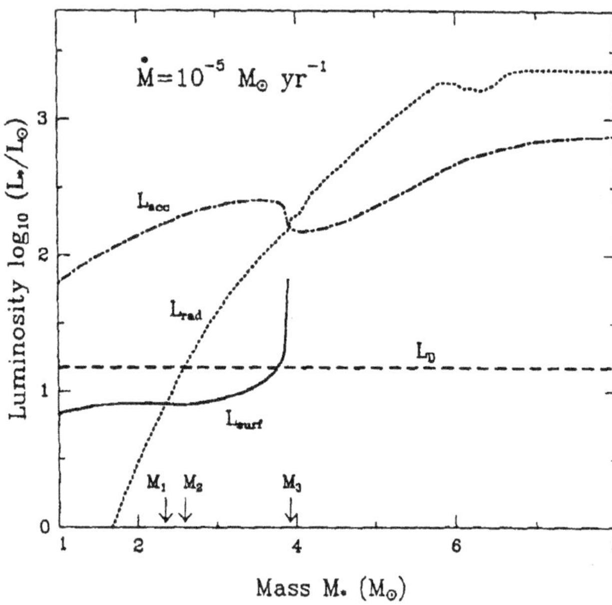

Fig. 12. The mass dependence of luminosities in protostars accreting at 10^{-5} $M_\odot$ yr $^{-1}$. The arrows on the horizontal axis indicate the critical masses where $L_{\rm rad}$ intersects $L_{\rm surf}$, $L_{\rm D}$, and $L_{\rm acc}$, respectively. At M_3, $L_{\rm surf}$ climbs steeply to join $L_{\rm rad}$ (from Palla & Stahler [41]).

Let us first apply Fig. 12 to the low mass T Tauri stars ($M_* <2$ $M_\odot$). According to the figure, such stars begin contracting with $L_{\rm surf}$ greater than $L_{\rm rad}$. In other words, they radiate more energy per unit time from their surfaces than can be supplied by photons from the interior. The excess luminosity must be carried by thermal convection. Thus, T Tauri stars begin their evolution as fully convective objects with nearly fixed surface temperatures, in the same manner described by Hayashi.

More massive stars, the Herbig Ae/Be stars ($M_* > 2\ M_\odot$), are not initially convective from surface cooling; a large fraction start the PMS contraction with $L_{\rm surf}$ less than $L_{\rm rad}$. In this case, more energy is transported per unit time through the deep interiors than can be radiated into space and the star is in a thermally unrelaxed state. In the H-R diagram, this star appears first with a surface luminosity and temperature that place it *below* the classical evolutionary tracks for that mass. During subsequent relaxation, the stellar radius, stellar luminosity, and surface temperature all increase, while the outer convections disappears. Then, the star joins onto the radiative portion of its classical track, which it then follows to the main sequence.

Not all the intermediate mass stars undergo thermal relaxation. If a star is sufficiently massive, it will have already relaxed prior to the end of protostellar accretion, i.e. before the star is optically visible. This condition is met when $L_{\rm rad}$ becomes larger than $L_{\rm acc}$. In Fig. 12 the critical mass is about 4 $M_\odot$. More massive Herbig Ae/Be stars appear immediately on their classical radiative tracks.

4.2 Stellar Structure Equations

Unlike the previous hydrodynamical protostellar phase, the PMS evolution of a star can be followed by models in hydrostatic and thermal equilibrium. These models are much easier to construct and less subject to numerical inaccuracies. The subject of stellar structure equations is well established and fully described in numerous textbooks. Here, I will present a brief description of the main equations needed for the calculation of PMS stellar models.

The *mechanical* stellar structure equations are simply a restatement of hydrostatic balance. If we neglect internal rotation, then a convenient spatial variable in a spherical star is the mass coordinate $M_{\rm r}$, defined by

$$M_r \equiv \int_0^{R_*} 4\pi r^2 \rho dr \ , \tag{30}$$

where ρ is the mass density of the gas. The function $M_{\rm r}$ represents the mass enclosed within a sphere of radius r. From eq. (30) we have immediately $\frac{dM_r}{dr} = 4\pi r^2 \rho$. The radius r then acts as a dependent variable, whose variation with M_r is governed by the inversion of this expression

$$\frac{dr}{dM_r} = \frac{1}{4\pi r^2 \rho} \ . \tag{31}$$

The internal pressure is related to the total weight by

$$\frac{dP}{dr} = -\frac{G\rho M_r}{r^2} \ . \tag{32}$$

Dividing (31) by (32) we derive the alternate form

$$\frac{dP}{dM_r} = -\frac{G M_r}{4\pi r^4} \ , \tag{33}$$

which again utilizes M_r as the independent variable. The pressure itself obeys the equation of state for an ideal gas:

$$P = \frac{\rho}{\mu} \mathcal{R} T \, . \tag{34}$$

Here the mean molecular weight μ depends on the state of ionization and dissociation of the gas, and is therefore a function of ρ and T. One may either calculate this function *ab initio*, by invoking statistical equilibrium, or obtain it from prior tabulations.

We turn next to the *thermal* stellar structure equations. Since the protostar's interior is highly opaque, the transport of radiation is governed by the diffusion equation

$$T^3 \frac{dT}{dM_r} = - \frac{3 \, \kappa \, L_{\rm int}}{256 \, \pi^2 \, \sigma_B \, r^4} \, . \tag{35}$$

The Rosseland mean opacity κ is again a function of ρ and T. Like μ, it is available in numerical form.

Finally, we consider the spatial variation of $L_{\rm int}$. Recall that the latter is the surface integral over a spherical shell of $|F_{\rm rad}|$, where the flux points radially outward. In general, a fluid element gains heat either by being irradiated externally or from internal nuclear reactions. Let $\epsilon \, (\rho, T)$ represent the rate of nuclear energy release per unit mass. Both this rate and $F_{\rm rad}$ enter as source terms in the *heat equation*:

$$\rho T \frac{ds}{dt} = \rho \epsilon - \nabla \cdot F_{\rm rad} \, , \tag{36}$$

where s is the entropy per unit mass of the fluid. For a spherical star, it is convenient to recast this relation as

$$\frac{dL_{\rm int}}{dM_r} = \epsilon - T \frac{ds}{dt} \, . \tag{37}$$

Equations (31), (33), (35), and (37) are the desired stellar structure equations for the dependent variables r, P, T, and $L_{\rm int}$. They must be supplemented by the equation of state (34) and by knowledge of μ, κ, ϵ, and s as functions of ρ and T. In practice, tabulations of the entropy utilize the Second Law of thermodynamics:

$$T \, \Delta s = c_v \, \Delta T - P \, \Delta\rho/\rho^2 \, , \tag{38}$$

which governs small changes of the state variables. Here, c_v is the specific heat at constant volume. Knowing this latter quantity (or, equivalently, the specific internal energy) as a function of ρ and T allows one to integrate eq. (38) in the $\rho - T$ plane and numerically obtain the function $s(\rho, T)$. Note that a purely monatomic gas would have $c_v = (3/2)\mathcal{R}/\mu$, where μ is now a constant. In this case, we may integrate eq. (38) analytically to obtain

$$s = \frac{\mathcal{R}}{\mu} \ln\left(\frac{T^{3/2}}{\rho}\right) + s_\circ \, , \tag{39}$$

where s_o is an arbitrary constant. Equation (39) is often a useful approximation in stellar interiors, where the gas is fully ionized and $\mu = 0.61$.

Solution of the stellar structure equations requires specification of four boundary conditions. Two of these are the statements that both $r(M_r)$ and $L_{\rm int}(M_r)$ vanish at the center of the configuration:

$$M = 0: \quad r(0) = 0 \quad L_{\rm int}(0) = 0 \,. \tag{40}$$

A third condition is that $P(M_r)$ must equal the appropriate photospheric value when $M_r = M_*$:

$$M = M_*: \quad P_{\rm e} = \frac{2}{3}\frac{GM_*}{R_*^2\kappa} \,. \tag{41}$$

The fourth boundary condition concerns the surface value of the temperature and its relation to the luminosity. For a PMS star, this relation is the standard photospheric one given by eq. (4):

$$M = M_*: \quad L = 4\pi R_*^2 \sigma T_{\rm eff}^4 \,. \tag{42}$$

The full four stellar structure equations can now be solved as a two-point boundary value problem, in which the values of the pressure and entropy are guessed at the center, while the stellar radius and luminosity are guessed at the surface. One can integrate the equations from both directions to an interior fitting point, where the four guesses must be changed until all variables match.

The four basic ingredients required by the stellar models are the equation of state, opacity, treatment of convection and the nuclear generation rate. Each of these physical processes would require a lengthy discussion, which is beyond the scope of this chapter. Here, a brief summary of some basic aspects is outlined.

4.3 Equation of State

The importance of using a good equation of state (EOS) for stellar models has been discussed by Däppen [33] and by D'Antona [29] for the case of PMS stars. The task is to develop a formalism for the computation of the EOS valid from the conditions of the photosphere, where the temperature is typically a few 10^3 K and the density some 10^{-7} g cm^{-3}, to the center of a star where $T \sim 10^7$ K and $\rho \sim 10^2$ g cm^{-3}. As a first approximation, PMS stars, because of contraction and of the relatively low density, should follow the ideal gas law. However, the Saha formula is just an approximation that ceases to hold as soon as a small degree of degeneracy or other non-ideal gas effects are present. Mazzitelli [64] has discussed in detail the implications of including non-ideal gas effects on the hydrogen and helium ionization. An important result is that because of the decrease of the value of the adiabatic gradient in regions of partial ionization, stellar models are characterized by *lower* effective temperatures. The most recent efforts to derive reliable EOS for astrophysical use include Mihalas, Hummer & Däppen [68], as part of the international "Opacity Project" (OP), and Iglesias & Rogers [49], as part of the opacity project at Livermore (OPAL). A clear explanation of the physical approach behind the two computations and comparisons between the results are presented by Däppen [33].

4.4 Opacity

There has been a dramatic improvement in the calculation of the Rosseland mean opacity of gas mixtures over the last ten years. Prior to 1990, the only existing set of opacities were provided by the Los Alamos group, both at high ($T \gtrsim 10^4$ K) and low-temperatures ($T \lesssim 10^4$ K) (Huebner et al. [48]; Cox & Tabor [28]). In the latter regime, an important contribution has been done by Alexander and collaborators who considered the presence of molecules and grains in the computations (Alexander et al. [2]; Alexander & Ferguson [1]). The physical processes considered in the calculations include continuum sources of opacity due to 14 elements (H, H^-, H_2^+, H_2^-, He, plus heavy elements); atomic and molecular lines (about 60 million transitions); electronic and Rayleigh scattering due to H, H_2 and He; absorption and scattering of grains with radius between 0.005 and 0.25 μm.

At temperatures above 8000 K, both the OPAL and OP projects provide extended tabulations of the opacity, including free-free, bound-bound and bound-free transitions for many elements from hydrogen to iron. The motivation for these calculations is interesting. In 1982, Simon [78] made a plea for the re-examination of the heavy element contribution to opacities. He showed that an increase in the opacity from elements heavier than helium by factors of two or three in the envelope regions, would resolve a number of outstanding problems in the theory of pulsating stars. Two groups, OPAL and OP, responded to the plea and performed calculations that showed that indeed a better treatment of atomic physics led to significant opacity enhancements that helped to solve the problem of (Beta) Cepheid pulsation (Iglesias & Rogers [49]; see also Seaton [76] for a review). The opacity tables are constantly updated, as improvements in atomic physics are obtained (Rogers & Iglesias [72]; Iglesias & Rogers [50]; Seaton et al. [77]). As an illustration of the properties of the Rosseland mean opacity at selected densities, Fig. 13 shows the behavior of the opacity at low- and high-temperatures for different sets of tables. In the right panel, the peak at $T \sim 10^5$ K, called the *Z-bump*, visualizes the impact of the OPAL calculations on the behavior of the opacity in the inner regions. This feature stems from bound-bound transitions involving the inner electrons of heavier metals, principally iron. The main peak at $T \sim 10^4$ K is due to the photoionization of hydrogen. At lower temperatures, the opacity falls steeply because too few photons in the Planckian distribution have energies exceeding the ionization threshold of

13.6 eV. The decline in opacity with falling temperatures halts temporarily at near 2000 K, where the dominant process is now bound-bound transitions in molecules (left panel).

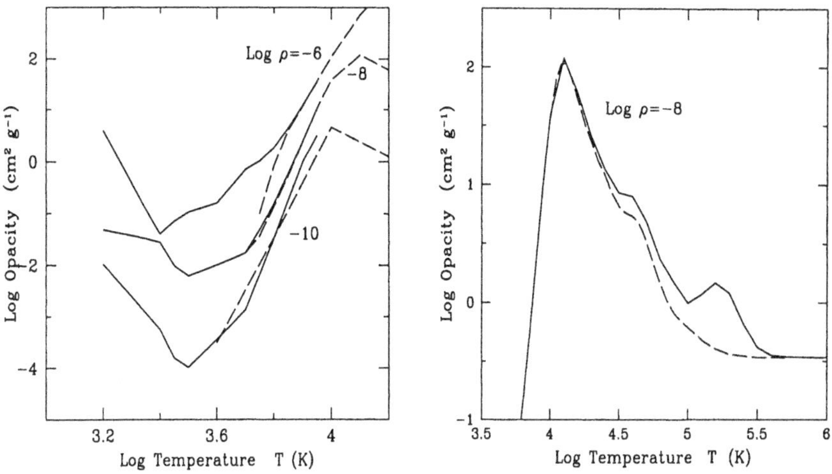

Fig. 13. Rosseland mean opacities for a gas of solar composition. *Left panel*: comparison of the low-temperature opacities for different densities. The solid lines are from Alexander et al. [1] and the dashed lines from Rogers & Iglesias [72]. Note that the latter do *not* include the contributions from molecules. *Right panel*: high-temperature opacities for a density of 10^{-8} g cm^{-3}. Solid line is from Rogers & Iglesias [72] and the dashed line is from Cox & Tabor [28]. Note the enhancement of the opacity at T$\sim 10^5$ K (*Z-bump*).

Molecules play a dominant role in the atmospheres of cool dwarfs, from mid K to late M stars, and a mass from 0.6 $M_\odot$ to the hydrogen burning limit (0.075-0.085 $M_\odot$, depending on metallicity), and in the brown dwarf regime. In such stars, most of the hydrogen is locked in H_2 and most of the carbon in CO, with excess oxygen bound in molecules such as TiO, VO, and H_2O (e.g. Allard et al. [4]). These molecules can serve as useful diagnostics of the nature of the very low-mass stars. Opacities due to grains are also becoming available (e.g. Alexander et al. [3]). Grains can condensate in stellar atmospheres at temperatures below $\sim$2000 K, and their effect is to decrease the gas phase abundance of some important absorbers, such as TiO. The impact of grain opacity on the spectral distribution and atmosphere of cool stars is quite important, but a full understanding of the physics of grain condensation and absorption is still lacking.

4.5 Treatment of Convection

The standard treatment of turbulent convection is a phenomenological one, based on Prandtl's theory of turbulence with the introduction of a quantity called the *mixing length* (Böhm-Vitense [14]). Despite its limitations, the mixing length theory (MLT) gives a satisfactory description of the convective zone since this region is to a large extent very close to being adiabatic. In the MLT theory, the length l over which a convective bubble merges with its surroundings is assumed to be proportional to the pressure scale height H_P, $l = \alpha H_P$, where α is a free parameter and $H_P = -dr/dlnP$. The actual value of α is chosen as to fit the solar radius and effective temperature. In general, α=1.5 yields a good fit. However, the surface properties of stellar models are quite sensitive to the value of α adopted in the computations. Lowering (rising) α to 1 (2) instead of 1.5 reduces (increases) the values of the effective temperature by about 10% (i.e. several hundred degrees), especially along the Hayashi vertical track (e.g. Forestini [39]). Thus, the poor knowledge of stellar convection introduces considerable uncertainties in the location of the tracks in the H-R diagram. Also, the temperature at the base of the convection zone is very sensitive to the value of α. Therefore, important differences in the level of light elements depletion in the convective envelopes result from different choices of the mixing length parameter.

An alternative theory to the MLT has been presented by Canuto & Mazzitelli ([19], CM) and Canuto, Goldman & Mazzitelli [20]. Unlike MLT which considers a single element of turbulence ("one-eddy" description), the CM convection model takes into account the full spectrum of the eddy distribution. Also, the mixing length is chosen to be the harmonic mean between the distance z of the layer from the top and the distance from the bottom of the convective region. The properties of the Sun can then be fitted to high precision (less than 1%, as demanded by helioseismological constraints) by introducing a small amount of overshooting from the surface convective layer and the prescription for the scale length becomes $l = z + \beta H_P$. A full description of the CM convection model, together with a discussion of the reasons to adopt such a model rather than the MLT, can be found in D'Antona & Mazzitelli [32].

4.6 Nuclear Reactions

Nucleosynthesis during the PMS phase is limited to the light elements. In addition to the destruction of deuterium, the nuclear burning of ^{6}Li, ^{7}Li, ^{9}Be, and ^{3}He takes place when the central temperature in solar-type stars reaches values of 2×10^6 K, 3×10^6 K, 3.5×10^6 K, and 8×10^6 K. However, the relatively tiny abundances of these species render the effect energetically and structurally unimportant. For example, the interstellar lithium abundance is [Li/H]$\sim10^{-9}$: in order to halt PMS contraction, the abundance should be enhanced by a factor 10^5 or more over the observed ones. On the other hand, the depletion history of lithium in low-mass stars can be followed in detail to compare with observations and to test the internal structure of stars. A full discussion of lithium depletion can be found in Ventura et al. [85].

The nuclear network consists of the complete *pp* chain and the CNO cycle. Deuterium and ^{7}Li burning are considered separately, since they occur before the *pp* chain takes over. The computation of the energy generation requires the knowledge of the nuclear cross sections. These are tabulated by Caughlan & Fowler [22], while weak and intermediate screening factors can be computed using the formalism of Graboske et al. [41]. Screening is also important for D and ^{7}Li burning reactions, especially for very low-mass stars and brown dwarfs. The relative concentration of the nuclear species follows the solar composition (e.g. Grevesse & Sauval[42]) with a metal abundance Z=0.019. The location of PMS tracks is not sensitive to the exact value of the metallicity; however, the amount of ^{7}Li depletion is strongly affected by Z, increasing for larger values of Z (D'Antona & Mazzitelli [30]).

5 The H-R Diagram

The results of the calculations of PMS evolution incorporating the protostellar initial conditions and the input physics discussed in Sect. 4 are shown in the H-R diagram of Fig. 14 for stars from 0.1 to 6.0 $M_\odot$. The evolutionary tracks have been computed using the opacities of Alexander et al. [1] at T<8000 K and of Rogers & Iglesias [72] at higher temperatures. Convection has been treated using MLT with α=1.5. From Fig. 14, we can see that while low-mass stars descend from the birthline on convective paths, stars more massive than $\sim$4 $M_\odot$ move to the left on radiative ones. Stars between 2 and 4 $M_\odot$ undergo thermal relaxation, looping back slightly behind the birthline and recrossing it before pursuing horizontal tracks.

Compared to the classical set of PMS tracks, those shown in Fig. 14 occupy a much more limited portion of the diagram. The reason is that the older tracks correspond to stellar radii too large to have attained during protostellar accretion. Thus, the starting luminosities are much lower than previously envisioned. In addition, the surface temperature of these stars begin *higher.* For example, a 5 $M_\odot$ star starts with $T_{\rm eff}$=12000 K in contrast to the value of 4400 K obtained by Iben [25]. Notice also that the birthline intersects the main sequence at a mass $M_* \sim$8 $M_\odot$: this point represents the critical stellar configuration in which hydrogen burning has stopped gravitational contraction while the star is still growing in mass (see Fig. 6). *More massive stars, therefore, have no PMS phase at all, but appear directly on the main sequence once they are optically revealed.*

An important result of the new calculations is that observed PMS stars should not be located to the right of the birthline. For both T Tauri and Herbig Ae/Be stars the agreement is excellent, despite the rather strong assumption of a constant mass accretion rate of 10^{-5} $M_\odot$ yr^{-1} used in the models.

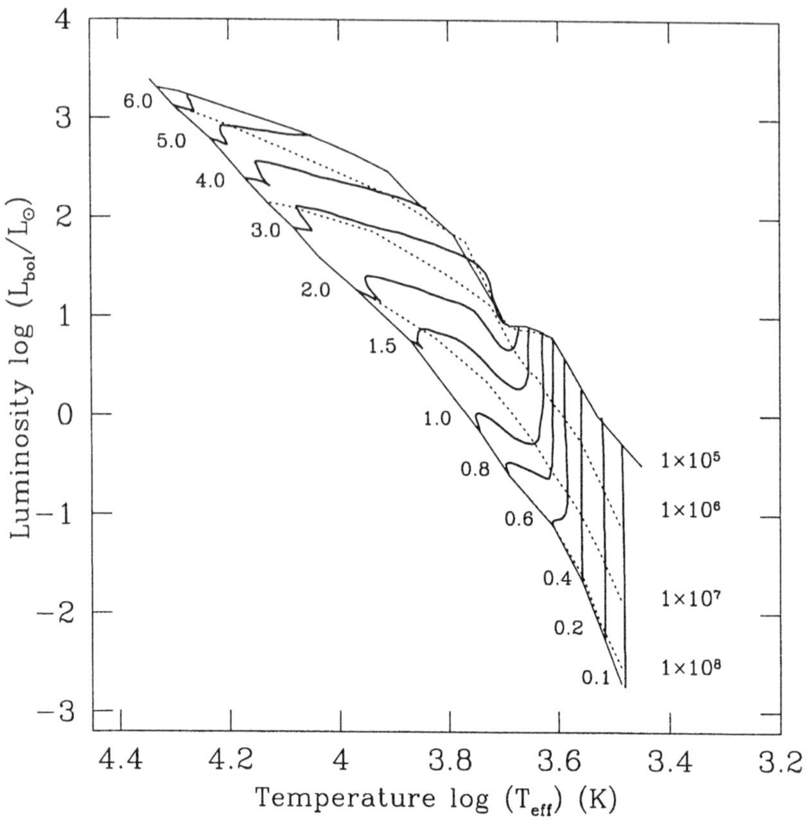

Fig. 14. Evolutionary tracks in the H-R diagram for low- and intermediate-mass stars. Each track is labeled by the corresponding mass, in solar units. Selected isochrones are shown by the dotted lines. For each track, the evolution starts at the birthline (not shown), and ends at the ZAMS, as indicated. (From Palla & Stahler [42])

This fact indicates that a limited dispersion in the properties of molecular cloud cores (which determine the magnitude of $\dot{M}_{\rm acc}$) somehow produces an impressive range of stellar masses. The case for even more massive stars is complicated by the difficulty in determining the appropriate stellar parameters to place them in the H-R diagram

Finally, another important result is the drastic reduction in age obtained for Herbig Ae/Be stars. The same is true, of course, for T Tauri stars, but the quantitative impact is minor. Even a large drop in the initial radius of a typical T Tauri star hardly changes its total contraction age, since t_{KH} scales as R_*^{-3}. For the Herbig stars, however, eq. (5) implies that t_{KH} only increases as $R_*^{-1/2}$. The age reduction is far more significant for masses well below the upper cutoff at $\approx 8\ M_\odot$.

Other key features of the evolutionary tracks of Fig. 14 are listed in Table 1. For each low-mass star, the second column gives the Δt_D, the duration of deuterium burning. This interval is measured to the point where $[D/H]$ falls to 0.1 its value on the birthline, but only for those masses where the initial abundance is at least 0.1 of interstellar. The third column lists $t_{\rm rad}$, the time at which a radiatively stable core first appears in fully convective models. Here, $t = 0$ refers to the star's appearance on the birthline. Finally, the last column gives $t_{\rm ZAMS}$, the total time required for each star to settle onto the zero-age main sequence. The latter is arbitrarily set at the point where the luminosity released through gravitational contraction falls to 3% of the total.

Table 1. Characteristic Times of Pre-Main-Sequence Evolution

Mass $(M_\odot)$	$\Delta t_{\rm D}$ (yr)	$t_{\rm rad}$ (yr)	$t_{\rm ZAMS}$ (yr)
0.1	1.5×10^6	–	3.7×10^8
0.2	7.0×10^5	–	2.4×10^8
0.4	3.0×10^5	1.1×10^7	1.1×10^8
0.6	2.0×10^5	5.5×10^6	8.5×10^7
0.8	1.5×10^4	2.5×10^6	5.2×10^7
1.0	–	1.4×10^6	3.2×10^7
2.0	–	–	8.5×10^6
3.0	–	–	2.0×10^6
4.0	–	–	8.2×10^5
5.0	–	–	2.3×10^5
6.0	–	–	4.0×10^4

5.1 Low-Mass Stars: $0.1 \lesssim M_*/M_\odot \lesssim 2$

The largest departure from the standard theory of PMS stars concerns the history of deuterium burning. Salpeter [73] was the first to recognize that deuterium ignition via the reaction ^{2}H(p,γ)^{3}He at a temperature of about 1×10^6 K could slow down contraction or, for the least massive stars, stops it entirely for a relatively brief time. This effect was confirmed by the detailed calculations of Grossman & Graboske [43] and Mazzitelli & Moretti [65] who introduced the concept of a "deuterium main sequence" to denote the curve in the H-R diagram crossed by the PMS tracks at the point of ignition. However, Stahler [46] has shown that this concept is obsolete, since deuterium burning occurs already during the accretion phase for stars with masses above a few tenths of a solar mass. Thus, the locus of the deuterium main-sequence is nothing else than the birthline shown in Fig. 7.

T Tauri stars, having consumed most of the available deuterium during the accretion phase, begin the contraction phase by simply descending along the vertical portion of the Hayashi track. The same is true for a Herbig Ae/Be star, which begins with a severely depleted deuterium supply. While this supply lasts, the outermost layers remain convectively unstable. But the luminosity provided by deuterium burning, $L_{\rm D}$, is now so minor compared to $L_{\rm rad}$ that even this limited convection vanishes by the time the star has joined the classical radiative portion of its evolutionary track. Finally, for stars of mass smaller than ~0.3 $M_\odot$, protostar theory still implies deuterium burning during the PMS phase, while for masses as low as M_*=0.018 $M_\odot$ degeneracy effects set in and prevent nuclear burning (D'Antona & Mazzitelli [31]).

It is convenient to consider separately those stars which contract as fully convective objects from those with radiative interiors. Referring again to Fig. 12, the distinction can be established through comparison of the two luminosities $L_{\rm surf}$ and $L_{\rm rad}$. Stars with $M_* < M_1$ are fully convective due to surface cooling as soon they appear as optically visible objects. The value of M_1 of 2.35 $M_\odot$ is coincindentally close to the mass, M_2, if 2.44 $M_\odot$ at which the radiative barrier first appears. The fact that $M_1 \approx M_2$ means that stars that were fully convective as protostars from deuterium burning remain so, at least for a time, following the end of accretion.

Figure 15 illustrates the first 3×10^6 yr in the PMS evolution of stars with masses of 1.5 and 2.0 $M_\odot$. Since both stars are fully convective at the beginning, they follow paths on the Hayashi tracks. However, their initial radii are much smaller, so that the convective phases are considerably shorter. In particular, the 2 $M_\odot$ star, with a mass close to the critical value M_1, contracts for only 1.1×10^4 yr before a radiative core appears at its center. The figure shows that the stellar radius actually swells during the spread of the radiative region: this swelling is a sign of nonhomologous contraction.

The nature of contraction is further shown by the evolution of the specific entropy in Fig. 16 for the star with 2.0 $M_\odot$. The star begins with the uniform entropy distributions enforced by convective mixing, apart for the sharp drop in entropy near the surface, where hydrogen and helium are recombining. The spreading of the radiative regions, in which the entropy increases outward, is also seen in both models. As the star radiates heat from the surface, its *average* interior entropy must drop. For lower mass stars, the decline within the radiative core is much faster than that in the convective exterior. For the 2 $M_\odot$ star, in which the radiative region has spread farther, the exterior entropy actually increases. In both cases, a portion of the heat lost from the deep, radiative interior is being transferred to the convective surface layers. This surface entropy rise is responsible for the swelling of the 2 $M_\odot$ mass star.

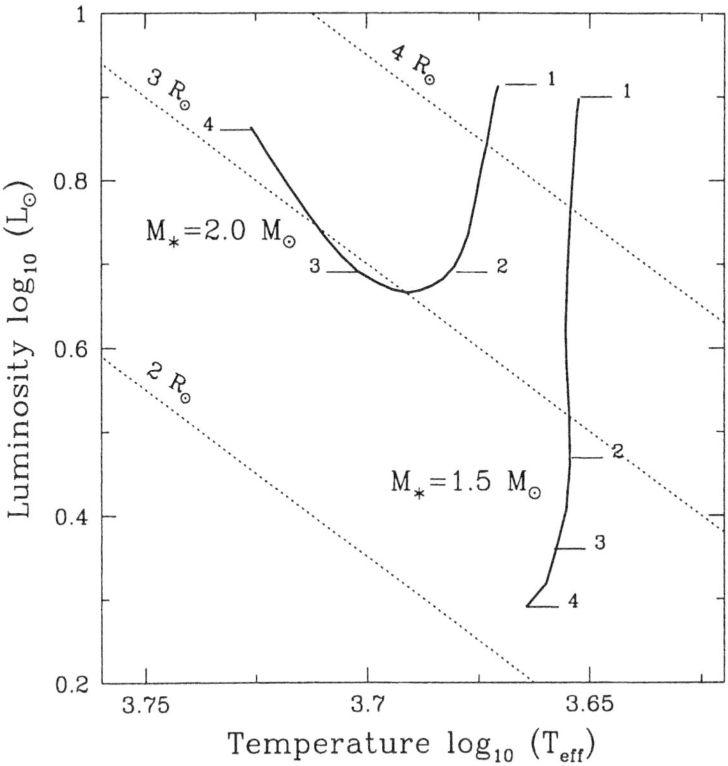

Fig. 15. Early evolution in the H-R diagram of stars with $M_* = 1.5$ and 2.0 $M_\odot$. The labeled tick marks on each curve correspond to times of 0, 1, 2 and 3×10^6 yr after the end of protostellar accretion. The thin dashed lines are curves of constant radius. (From Palla & Stahler [41])

The evolutionary trend seen for this, and also for the lower mass stars, is identical to that found in traditional calculations. In particular, the entropy distribution is undisturbed by the fusion of residual deuterium. As shown in Fig. 11, the initial concentration for stars more massive than $\sim$0.5 $M_\odot$ is so low that it is quickly consumed. Since the deuterium luminosity is initially close to the steady state value in eq. (23), the time for the exhaustion of the fuel, Δt_D, is

$$\Delta t_D = \frac{f_D\,\Delta M_D\,[D/H]\,X\,Q}{m_H\,L_D} = \frac{f_D\,\Delta M_D}{\dot{M}}\,. \qquad (43)$$

In this equation, ΔM_D represents the stellar mass contained in the convective region. It turns out that Δt_D = 30 yr for 1.5 $M_\odot$ and only 1 yr for 2.0 $M_\odot$. These exceedingly brief times serve to emphasize the point that PMS convection is driven in these stars by surface cooling, rather than central heating. For lower mass stars, however, the times for deuterium exhaustion

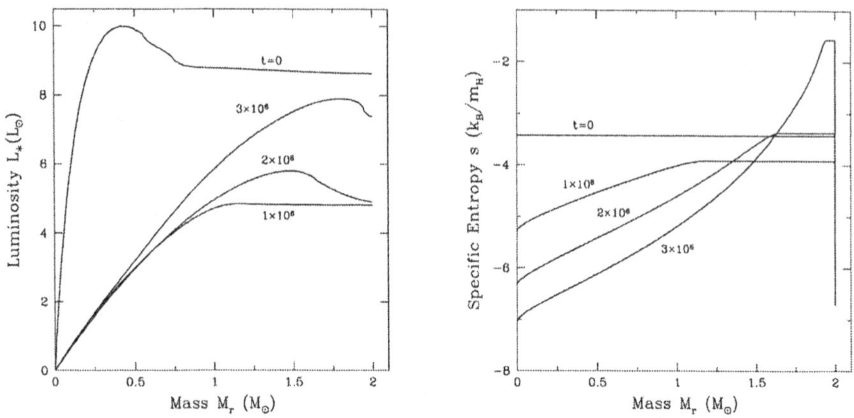

Fig. 16. Early evolution of the entropy (right panel) and luminosity (left panel) profiles for M_* =2.0 $M_\odot$. The curves show the specific entropy and luminosity as a function of the internal mass coordinate. Each curve is labeled by the corresponding evolutionary time. Specific entropy is measured in units of k_B/m_H, with the zero point arbitrarily set at $T = 2.05 \times 10^5$K and $\rho = 5.16$g cm $^{-3}$. (From Palla & Stahler 1993)

are longer, as listed in Table 1. Stars below ~0.5 $M_\odot$ have almost the full content of interstellar deuterium initially and take approximately 10^6 yr to burn it completely.

The rapid disappearance of deuterium is reflected in the stars' interior luminosity. The behavior during the first 3×10^6 yr is shown in the left panel of Fig. 16 for the 2.0 $M_\odot$ star. At $t = 0$, the luminosity rises rapidly from the center and attains a value slightly less than ~12 $L_\odot$, generated by the steady-state burning of deuterium. Then, the absorption of this energy by the overlying layers of gas causes the decline at larger mass shells. However, after a period Δt_D, the luminosity arises from gravitational contraction alone and has a very different distribution. Since most of the interior is homologously contracting, the global loss of heat leads to a luminosity that monotonically rises toward the surface. Between $t = 1 \times 10^6$ and 3×10^6 yr, the surface luminosity rises despite the increase in stellar radius: this rise is due to the disappearance of the absorbing convective layer.

5.2 Intermediate-Mass Stars: $2 \lesssim M_*/M_\odot \lesssim 10$

For stars more massive than solar, the imprint of the prior accretion history persists much longer than in the low-mass, convective case. The reason is that the stellar interior is not thermally relaxed, and the stars must undergo a phase of global readjustment. The main results can be schematically summarized as follows.

• *Fully convective stars*: Stars in the range $1-2.5\ M_\odot$ are fully convective due to surface cooling as soon as they appear as optically visible objects. The evolution is undisturbed by the fusion of residual deuterium: its concentration is so low, that it is quickly consumed (on a time scale of few years). Stars contract along the Hayashi track, but their path is much reduced (see the case for $M_* = 2\ M_\odot$ in Fig. 14).

• *Partially convective stars*: Stars in the range $2.5-4\ M_\odot$ undergo thermal relaxation and nonhomologous contraction. When they are first optically visible, they are *underluminous*: i.e., the evolutionary track begins at low luminosity and then moves up to join the radiative portion. The Hayashi phase is skipped entirely. In the H-R diagram, shown in Fig. 17 for the case of a 3.5 $M_\odot$ star, the unrelaxed star first appears below the classical evolutionary track for that mass. At this early epoch, the star has a relatively shallow convection zone. During subsequent relaxation, the stellar radius, surface luminosity and temperature all increase, while the outer convection disappears. The variation of the distribution of the internal luminosity at different times is also shown in Fig. 17. Initially, the function $L_{\rm r}(M_{\rm r})$, the luminosity crossing any interior shell enclosing a mass $M_{\rm r}$, rises from zero at the center, peaks somewhere in the interior, then falls again to the relatively low surface values $L_{\rm surf}$. Such odd luminosity profile cannot be maintained indefinetely. According to the heat eq. (36), matter between the star's center and the shell where $L_{\rm r}$ peaks is losing heat ($ds/dt < 0$), while the gas between this shell and the surface is gaining heat. As shown in Fig. 17, this internal heat transfer, achieved entirely through radiative diffusion, causes the peak of $L_{\rm r}$ to move outward with an attendant rise in the surface luminosity. The approximate duration of this *thermal relaxation* is given by the Kelvin-Helmholtz time with $L_{\rm surf}$ replaced by the internal luminosity $L_{\rm rad}$ in eq. (9).

The luminosity in stars of mass between $2.5-4\ M_\odot$ stems almost entirely from gravitational contraction. This contraction, however, is initially quite different from that envisioned in the classical PMS theory. That interior portion which is losing heat indeed shrinks, but the transfer of this energy actually expands the outer region and creates a net increase of the stellar radius. This *nonhomologous contraction* continues until $L_{\rm r}(M_{\rm r})$ is everywhere increasing, by which point the star as a whole is shrinking, as in the classical model.

• *Fully radiative stars*: Stars more massive than 4 $M_\odot$ appear immediately on the radiative portion of the classical track, and begin to contract homologously under their own gravity. As the star contracts, the average interior luminosity must rise as $R_*^{-1/2}$, according to eq. (5). Such rise is evident in the evolutionary tracks of Fig. 14. The specific entropy decreases with time in all interior mass shells, a sign of homologous contraction. At any one time, the interior is radiative, i.e. the entropy is increasing outward, except for a very thin surface layer.

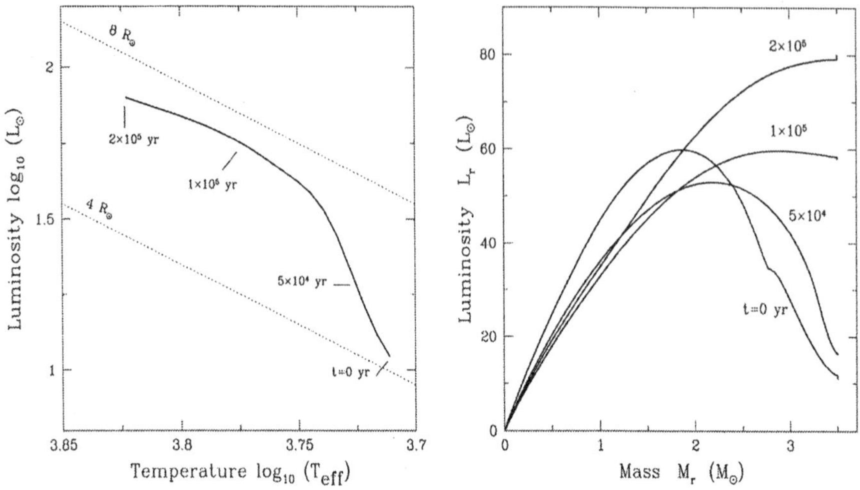

Fig. 17. *Left panel*: Early evolution in the H-R diagram of a 3.5 $M_\odot$ star, covering the phase of thermal relaxation. The thin dashed lines are cruves of constant radius. *Right panel*: The evolution of the internal luminosity during the thermal relaxation process. Note the shift of the peak in luminosity toward the stellar surface as a result of nonhomologous contraction. (Adapted from Palla & Stahler [41])

5.3 Massive Stars: $M_* \gtrsim 10 M_\odot$

The process of gravitational contraction during the accretion phase ends when the nuclear luminosity due to the CNO cycle becomes dominant. There is no precise way to specify the arrival of a protostar at the main sequence, since the models are never identical to ZAMS stars of the same mass. For example, at M_*=8 $M_\odot$, the nuclear luminosity still accounts for 75% of the total luminosity, with the remainder provided by gravitational settling in the outer layers. However, we know that the stellar radius must reach a minimum if accretion were to increase indefinetely. From Fig. 6, the value of the minimum R_* is 3.1 $R_\odot$. This radius is reached on the theoretical ZAMS at $M_* \approx 8$ $M_\odot$. Thus, we may take this mass as the value where the protostar has joined the main sequence. The consequence is obvious: there is no longer any PMS phase for stars more massive than this value. They can only be observed as infrared accreting protostars or as main-sequence or post-main sequence objects.

An implication of the evolution described so far is that most stars should form with nearly the same accretion rate, as supported by the distribution of known PMS star in the H-R diagram. One possible objection to this hypothesis is that stars forming at modest rates might complete their entire main-sequence evolution *before* reaching a sufficiently high mass. However, with an $\dot{M}_{acc}$ of 10^{-5} $M_\odot$ yr^{-1}, the accretion and nuclear lifetimes become equal only at a stellar mass of 45 $M_\odot$, while with $\dot{M}_{acc}$=10^{-4} $M_\odot$ yr^{-1} the critical mass is higher than 100 M$_\odot$ (Schaller et al. [74]). These figures suggest that even high-mass stars form with accretion rates in the range considered above and do not require extreme conditions. This possibility has been considered by Beech & Mitalas [7] and Bernasconi & Maeder [8] who studied the formation and evolution of stars up to ~100 M$_\odot$ under the accretion scenario. The resulting birthlines are displayed in Fig. 18. One of the interesting findings is the definition of an upper stellar birthline as an extension above ~10 M$_\odot$ of that shown in Figs. 7 and 14. A massive star then begins its main-sequence evolution from these initial conditions that differ substantially from those at the classical ZAMS. In particular, they have lower effective temperatures and a reduced extent of the convective core.

However, accretion cannot proceed indefinetely because of the mechanical effect of the protostar's radiation on the infalling envelope. The accretion shock at the stellar and disk surfaces produces photons at optical and ultraviolet wavelengths. These are absorbed by the envelope's dust grains and reradiated into the infrared. Trapping of this infrared radiation creates thermal pressure that retards infall. The stellar luminosity also raises the envelope's temperature until the grains sublimate at the dust destruction front. The direct impact of short-wavelength photons on grains at this front provides an additional retarding force. Radiation pressure begins to become significant in protostars of intermediate mass. Infall cannot be stopped, however, as long as the accretion component dominates the total luminosity. Any retardation of the flow diminishes $\dot{M}_{\rm acc}$, which in turn lowers L_*. Once the luminosity falls, the flow picks up again. Numerical simulations of collapsing clouds with central stars display this effect. For stars in the proper mass range, $\dot{M}_{\rm acc}$ oscillates strongly without actually reversing (Yorke and Krügel [90]). The feedback between radiation pressure and infall is cut once $L_{\rm int}$ dominates L_*. Again, the transition occurs just where $t_{\rm KH}$ falls below $t_{\rm acc}$, *i.e.*, for protostars that have joined the main sequence. The internal luminosity is now supplied by hydrogen fusion rather than contraction, but still scales as $M_*^{11/2} R_*^{-1/2}$. Hence the retarding force on the envelope increases rapidly with mass. Within the context of spherical collapse, the result is inevitable and the condition of balance between the forces yields

$$\frac{L_*}{M_*} = \frac{4\pi c G}{\kappa} = 1 \times 10^3 \, \frac{L_\odot}{M_\odot} \,, \tag{44}$$

where it has been used a value of κ=10 cm^2 g^{-1} for the opacity. Plausible variations in grain composition could alter this value by as much as a factor of 3 in either direction (Draine & Lee [35]; Pollack et al. [71]). Thus, the critical ratio occurs at a mass near 10 M$_\odot$, implying that accretion is significantly affected for central objects of greater mass. Wolfire and Cassinelli [89] have analyzed the dynamics of envelopes falling onto stars with M_* exceeding 60 $M_\odot$. They found that infall was reversed unless $\dot{M}_{\rm acc}$ were arbitrarily increased to *at least* 10^{-3} $M_\odot$ yr^{-1} *and* grains were depleted in abundance

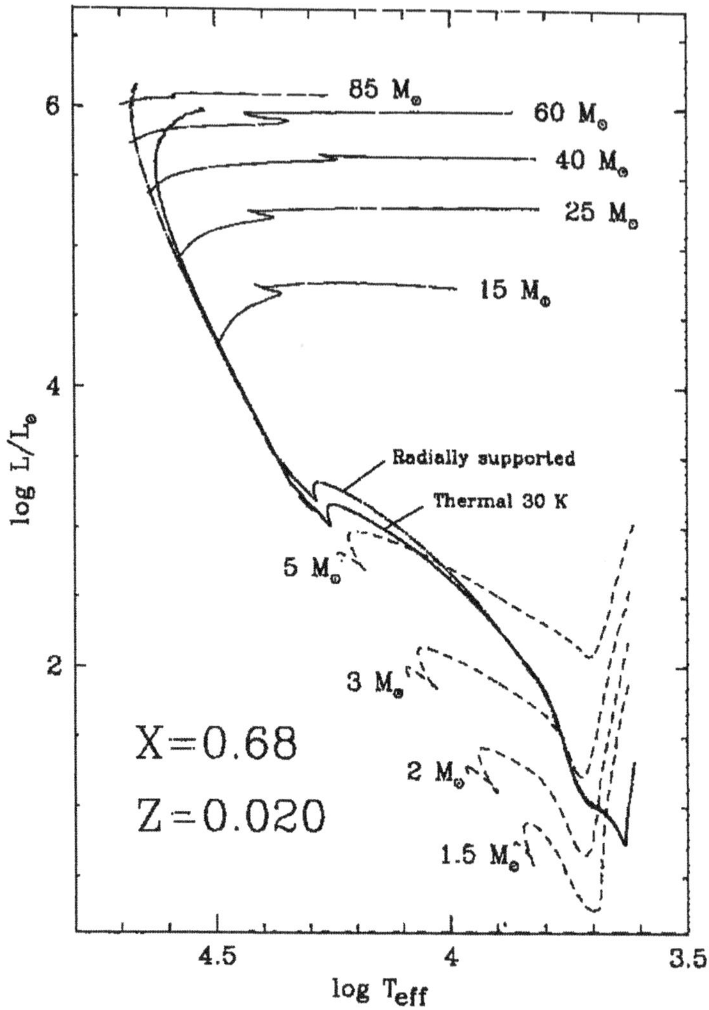

Fig. 18. The birthline for massive stars in the H-R diagram in the accretion scenario. The two birthlines refer to the case of constant accretion rate at $\dot{M}_{\rm acc} = 10^{-5}$ $M_\odot$ yr^{-1}, and to a time-dependent rate. The dashed lines are the standard PMS tracks for several masses computed from arbitrary initial conditions. Finally, the upper thin lines display the post-MS evolutionary tracks computed with mass loss for selected masses, as indicated. (From Bernasconi & Maeder [8])

by an order of magnitude. For earlier discussions on the problem of infall reversal, see Larson and Starrfield [56] and Kahn [51]. More recently, the possibility of forming massive stars by fast accretion has been quantitatively explored in a series of papers by Maeder and collaborators ([62], [63]).

There is little doubt that a massive star produces more than enough energy to expel its envelope. In fact, the inequality $t_{KH} < t_{acc}$ implies that the radiated luminosity can easily unbind the parent cloud before the star has gained appreciable mass. The real issue is the *efficiency* of energy transfer. Here, any departure of the infall from spherical symmetry is likely to play a major role (Nakano [69]). Unfortunately, a consistent theory which includes rotation and the simultaneous presence of infall and outflow is still unavailable, so that the formation of massive stars via accretion remains a mystery. Perhaps, one should invoke a completely different scenario (e.g. collision/coalescence models in dense clusters: Bonnell, Bate & Zinnecker [15], and Stahler, Palla & Ho [80]).

6 Tests to PMS Evolutionary Diagrams

The location of young stars in the H-R diagram is commonly used to estimate their masses and approximate ages. When young clusters or associations are studied, the information on the mass of individual members can yield direct information on the rate and efficiency of star formation and, eventually, on the initial mass function to compare to that observed locally. Also, the amount of depletion of the light elements due to nuclear activity in these early stages sheds light on the internal structure and on the mechanisms of angular momentum transport. It is important at this point to discuss the uncertainties that affect the computation of modern evolutionary tracks, such as those described earlier. The sources of uncertainty can be due to the uncertain knowledge of certain aspects of the input physics used in the theoretical models, or to the complex phenomena that accompany the evolution of young stars, such as the interaction with circumstellar disks and the accretion of residual matter, or the presence of companions and so on. In the following, I will first overview the uncertainties *intrinsic* to the models, and then discuss the effects of the environment.

6.1 Comparison of Tracks

Several recent studies have reconsidered the PMS evolution of stars with population I compositions over a wide mass range (e.g. D'Antona & Mazzitelli [31]; Forestini [39]; Swenson et al. [83]; Baraffe et al. [5]; Palla & Stahler [42]). The main motivation for these studies is the progress in the fields of the equation of state, radiative opacities and the treatment of convection, each of which greatly influences the surface properties, i.e. the observables, of PMS stars. Before summarizing the main properties of the new models, let first stress the following point: the average error in the determination of the observable quantities, L_{bol} and T_{eff}, is of the same order of magnitude as the theoretical ones. Typically, $\Delta T_{eff}=\pm 300$ K and $\Delta \log\ (L_*/L_\odot)=\pm 0.25$ dex, for stars observed in the nearby star forming regions (e.g. Strom & Strom [82]). On the other hand, the mean spacing between Hayashi tracks is of the order of ~130 K for M_*=0.5 $M_\odot$, and of only ~60 K for a 1 $M_\odot$ star (Forestini [39]). Thus, for very young stars, still on the vertical portion of the tracks, the mass assignment based on their position can lead to errors of a factor of 2.

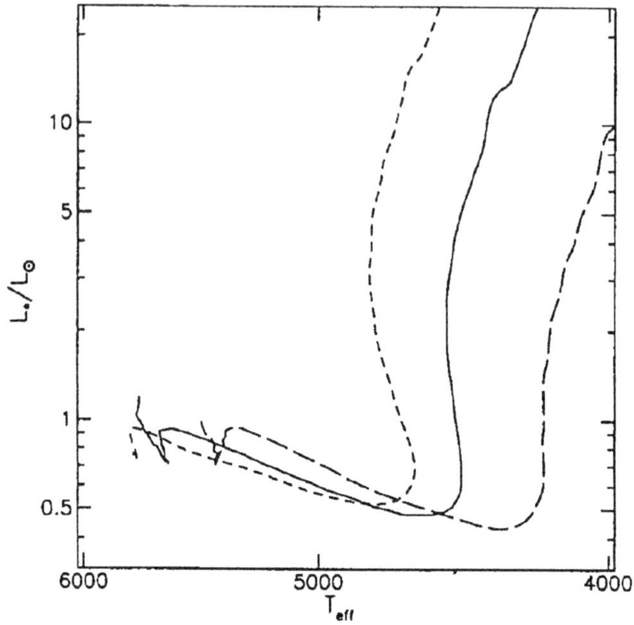

Fig. 19. Effects of the variation of the free parameter α in the mixing-length theory on the evolutionary tracks of a M_*=1 $M_\odot$ star. The solid line is for a model with α=1.5 which fits the Sun, while the other curves are for α=1 (redshifted) and 2 (blueshifted), respectively. (From Forestini [39])

The largest sources of uncertainty in the position of the tracks in the H-R diagram are due to the poor description of turbulent convection and the limited knowledge of low-temperature opacities. If the mixing-length theory is adopted, then the difference in effective temperature can amount to ~700 K for a variation of the free parameter α from 1 to 2 (the standard value for a M_*=1 $M_\odot$ star that fits the Sun is α=1.5). Such a variation is shown in Fig. 19 for a 1 $M_\odot$ star. The difference is thus comparable to the observational error. However, the discrepancy is greatly reduced if one follows the nonlocal treatment of convection developed by Canuto & Mazzitelli [19] and used in the PMS models of D'Antona & Mazzitelli [31] [32]. The variation in $T_{\rm eff}$ is lower than 3% at the top of the PMS tracks and becomes insignificant along the radiative portion. Note that the parameter a in the Canuto &

Mazzitelli theory is not a free parameter as in the mixing-lenght sense, but it is introduced only to allow for fine tunings in the calculations of the standard Sun and for internal comparisons. The important aspect to underline is that the value of α which fits the solar T_{eff} does not result in a Hayashi track having the same effective temperatures as for D'Antona & Mazzitelli (in the sense that the latter obtain *blueshifted* tracks). The differences however are within the theoretical uncertainties associated with the inaccuracy of the low temperature opacities, and therefore the two convection theories cannot be tested observationally in this way.

The overall level of uncertainty expected from some of the existing theoretical models can be appreciated by inspection of the evolutionary tracks shown in Fig. 20. Here, we compare the results of Palla & Stahler [42] to those of D'Antona & Mazzitelli [32]. The latter exhibit tracks with much greater curvature than those of Palla & Stahler, even in the fully convective case. The D'Antona & Mazzitelli models also have higher surface temperatures at the earliest ages. For masses between 0.4 and 1.5 $M_\odot$, the difference in temperatures are of about 400 K. Conversely, a young PMS stars which has a mass of 0.4 $M_\odot$ according to Palla & Stahler, would have only 0.2 $M_\odot$ using the other set of tracks. Equally severe are the discrepancies in age. The use of the birthline to set the initial time reduces the ages for intermediate masses. At the lowest masses, the D'Antona & Mazzitelli isochrones sag below the others. Thus, a star of given L_* and T_{eff} has a lower age, affecting the interpretation of the H-R diagram of young clusters.

An example is provided by the recent study of Luhman & Rieke [60] of the low-mass end of the IMF in L1495E in Taurus. As shown in Fig. 21, the results are quite different depending on the choice of the set of tracks: an object with a mass of 0.25 $M_\odot$ using D'Antona & Mazzitelli [31] tracks corresponds to a 0.5 M_* using Baraffe et al. [5] tracks. Thus, the shape of the IMF cannot be uniquely determined, and it is too premature to draw firm conclusions from observations of the low mass end of the stellar distribution with the current uncertainties in both evolutionary tracks and in the conversion from observational to theoretical quantities (see also the discussion in D'Antona & Mazzitelli [32]). We shall return to this important point later on when we discuss the global properties of clusters and associations.

Similarly, a deep CCD survey aimed at discovering the lowest mass PMS stars in Taurus has discovered nine new objects, all of them with spectral types later than M4 (Briceño et al. [17]). Their distribution in the H-R diagram is shown in Fig. 22. The comparison between various evolutionary models indicates that the D'Antona & Mazzitelli [31] and Burrows et al. [18] models tend to yield larger masses, whereas the smallest masses are obtained by the Baraffe et al. [6] tracks. However, if one assumes that $M_* = 0.075$ $M_\odot$ is the minimum mass for hydrogen burning at solar metallicity for ages greater than about 10^8 yr (Chabrier et al. [25]), then all models agree in predicting that T Tauri stars stars with spectral type M6 and later are young

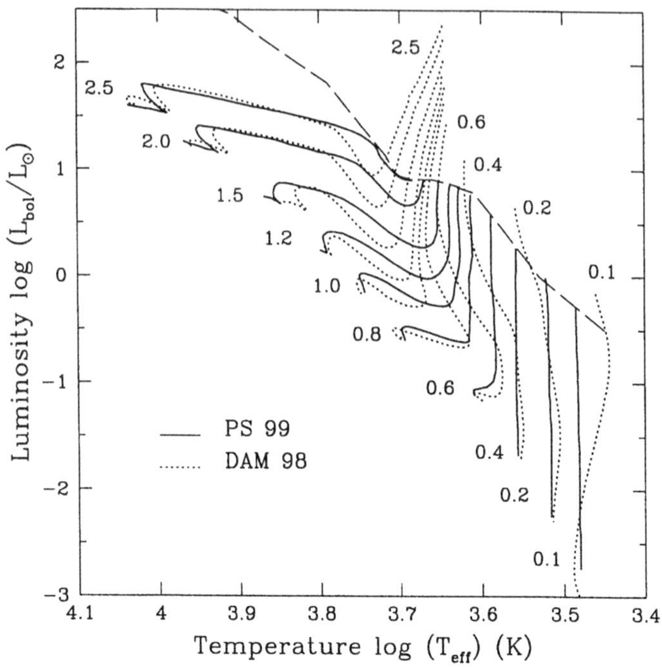

Fig. 20. Comparison of PMS evolutionary tracks of low-mass stars ($M_* \lesssim 2.5\ M_\odot$) computed by D'Antona & Mazzitelli (1998) using the Canuto-Mazzitelli prescription for convection (a =0.5), and by Palla & Stahler (1999) using MLT with α =1.5.

brown dwarfs. Finally, the ages derived from the models differ by not more than a factor of two, indicating that these low-mass stars are very young objects ($\lesssim 2 \times 10^6$yr), consistent with the average age of the more massive objects of the Taurus association.

6.2 The Low-Mass M Dwarfs and the End of the Main Sequence

In order to construct an H-R diagram, one should derive $T_{\rm eff}$ and $L_{\rm bol}$ from observable quantities. The derivation of $T_{\rm eff}$ relies on temperatures scales which are fairly well known for dwarfs between A and K types. For M dwarfs ($T_{\rm eff}$ <4000 K), differences between temperature scales are much larger (Allard et al. [4]). Below ~0.6 $M_\odot$, the observational mass-luminosity relation is determined by only ~18 stars, with mass determination at the level of 10%–20% accuracy (Henry & McCarthy [47]). The problem is that there are only two eclipsing binary systems, with accurately measured temperatures. Thus, it is difficult to plot a reliable H-R diagram for the low- and very low-mass stars and obtain good mass and age estimates.

Selection effects render the discovery of eclipses in intrinsically faint and small stars rather difficult. The two known eclipsing binary systems are YY

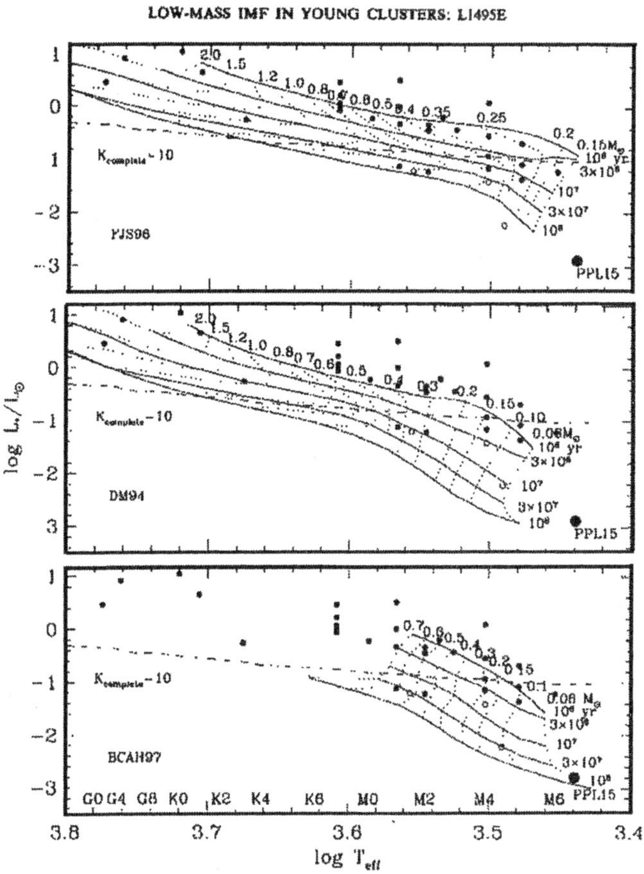

Fig. 21. H-R diagram of L1495E in Taurus–Auriga. The three panels show a comparison of different evolutionary tracks: FJS96 for Swenson et al. (1996), DM94 for D'Antona & Mazzitelli (1994), and BCAH for Baraffe et al. (1997). The solid dots are for members of the cluster, whereas the open circle are probable foreground stars. The dashed line indicates the completeness limit at dereddened K=10 mag. The position of the brown dwarf PPL 15 in the Pleiades is shown for comparison. (From Luhman & Rieke [60])

Gem and CM Dra with dynamical mass estimates of M(YY Gem A,B)= $0.62 \pm 0.03; 0.57 \pm 0.03$ $M_\odot$ (Leung & Schneider [59]) and M(CM Dra A,B)= $0.231 \pm 0.001; 0.214 \pm 0.001$ $M_\odot$ (Lacy [55]; Metcalfe et al. [66]). In YY Gem, the two components are nearly identical dMe stars in an almost circular orbit (e <0.01) viewed almost edge on ($i = 86.4°$). The eclipsing system CM Dra is also circular and edge on ($i = 89.8°$). CM Dra is currently the least massive main sequence eclipsing SB2 currently known and it provides the opportunity to test models of stellar interiors and atmospheres. Chabrier & Baraffe [23]

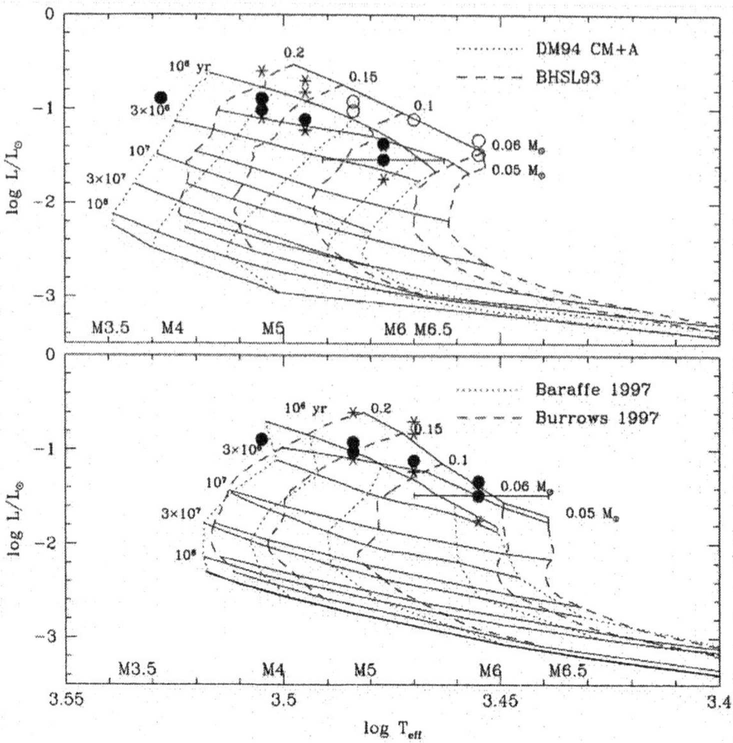

Fig. 22. The low-mass end of the HRD in Taurus-Auriga. The location of the the newly discovered T Tauri stars is displayed using two temperature scales: Stauffer et al. [81]) (*filled circles*) and Leggett et al. [57] (*open circles*). The two panels illustrate differences in the evolutionary tracks and isochrones computed by various groups, as indicated. (From Briceño et al. [17])

have presented evolutionary models that reproduce the observed masses, radii and luminosities of both binary systems.

Recently, Delfosse et al. [34] have discovered a third detached M dwarf eclipsing system: GJ 2609, a nearby (d =12.8 pc) M3.5V quadruple system, whose main component, GJ 2069A is the eclipsing binary. From accurate radial velocity orbit and light curve, Delfosse et al. derive individual masses of 0.433 ± 0.002 $M_\odot$ and 0.397 ± 0.002 $M_\odot$, respectively. Unfortunately, no estimates for the effective temperatures are available and we cannot use the stars as a test of the goodness of the tracks. From a comparison with the theoretical mass-luminosity relation of Chabrier & Baraffe [24], Delfosse et al. conclude that the components of GJ 2069A appear to be sub-luminous for their mass with respect to solar metallicity models (i.e. they fall slightly below the ZAMS), suggesting perhaps a metal-rich composition since an increase in the metal content shifts the ZAMS toward the blue. Another interesting case is the eclipsing system BW3 V38, discovered by the OGLE microlensing project (Maceroni & Rucinski [61]). With a period of only 0.2 day, this is the shortest period system known with indication of strongly distorted orbits. Although the parameters have not been fully derived, Maceroni & Rucinski conclude that the the properties of the components are intermediate between those of YY Gem and CM Dra.

Considering the stellar parameters of YY Gem and CM Dra, Habets & Heintze [45] assign a spectral type M1V to YY Gem and M4V to CM Dra. The conversion from spectral types to effective temperatures is rather delicate, especially for CM Dra, for which values vary between 3150 K (Lacy [55]) and 3350 K (Chabrier & Baraffe [23]) with a sensitivity on the stellar metallicity. Gizis [40] from optical spectroscopy finds a solar metallicity for CM Dra. Based on a direct comparison of synthetic and observed spectra in the near-IR and optical regimes, Viti et al. [86] derive values of T_{eff} in the range 3000-3300 K for various metallicities, with the latter value corresponding to the solar abundance. As for YY Gem, few estimates are available in the literature and they are restricted to values around 3750 K for both components (Habets & Henitze [45]). In both cases, the distance is well established by astrometric parallaxes.

The location of YY Gem and CM Dra in the H-R diagram is displayed in Fig. 23, along with other M dwarf stars without accurate mass measurements (Kirkpatrick et al. [53]; Leggett et al. [57]). The values adopted for YY Gem are $\log L = -1.28\ L_\odot$ and $T_{eff} = 3790 \pm 200$, while those for CM Dra (A,B) are $\log L = -2.30, -2.36\ L_\odot$ and $T_{eff} = 3200 \pm 150$K for both components. Thus, the mass estimate for YY Gem is $0.52 \pm 0.06\ M_\odot$, and that for CM Dra (A,B) is 0.18 ± 0.08. In either case, mass estimates tend to be lower than the dynamical values, but within the range of uncertainties.

6.3 Derivation of Observable Quantities from Evolutionary Tracks

It is often useful to derive directly observable quantities, such as colour indices and absolute magnitudes, from the stellar parameters computed by PMS models. For example, near-infrared observations of young embedded clusters can probe the stellar population down to very faint magnitudes and it is important to transform the K-band (or any other photometric band) completeness absolute magnitude into an estimate of the lowest mass detectable in the field. The discussion below follows closely the method described by Meyer ([67]).

First, one needs to convert the set of PMS evolutionary tracks from the theoretical H-R diagram ($\log L_\star$, $\log T_{eff}$) into the observed color-magnitude plane [M_K, (V–K)], where M_K is the absolute K magnitude. Useful relations are provided by the calibrations of Schmidt-Kaler [75], Bessell [9]; [10]), Bessell & Brett ([12]) and Koorneeff ([54]) for dwarf stars which can be used to compile tables of the bolometric corrections at the V-band (BC_V) and the (V–K) colour index as a function of the effective temperature. These relations

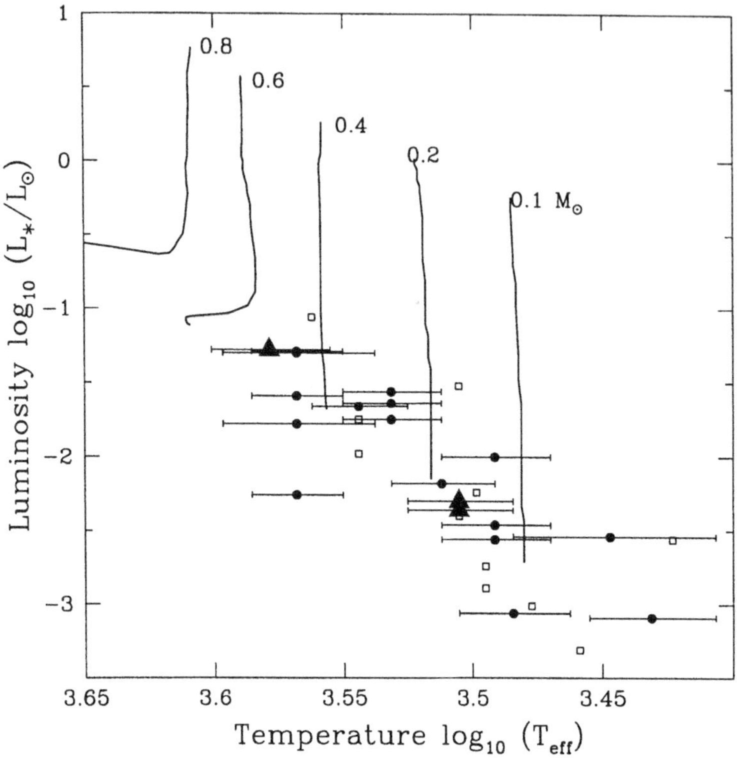

Fig. 23. H-R diagram for M dwarfs. The two eclipsing binaries, YY Gem and CM Dra, are shown by filled triangles (data from Viti et al. [86]; Bessell [11]). Data from Kirkpatrick et al. [53] are represented by open squares, and data from Leggett et al. [57] by filled circles. Several low-mass tracks from Palla & Stahler [42] are also shown for comparison.

are shown in Fig. 24. For each value of $L_\star$ and $T_{\rm eff}$, one can then compute M_K as:

$$M_K = 4.725 - 2.5\,\mathrm{Log}(L_\star/L_\odot) - BC_V(T_{\rm eff}) - (V-K)(T_{\rm eff}) \ , \qquad (45)$$

where 4.725 is the assumed absolute V-magnitude of the Sun. The colour-magnitude (V–K, M_K) diagram resulting from the transformation of the evolutionary tracks is shown in Fig. 25.

We can now derive for each stellar mass the run of the K-absolute magnitude with time. This is shown in the right panel of Fig. 25 for masses in the interval 2.5–0.1 $M_\odot$. The peak in M_K that appears for the more massive stars at Log(age)$\sim$ 6.2–7.4 is due to the transition from the convective to the radiative portion of the evolutionary tracks. We can see that in the range of ages (t$<$ 10 Myr) and *minimum masses* considered here, the M_K of a star of a given mass is a monotonically increasing function of time. For this reason,

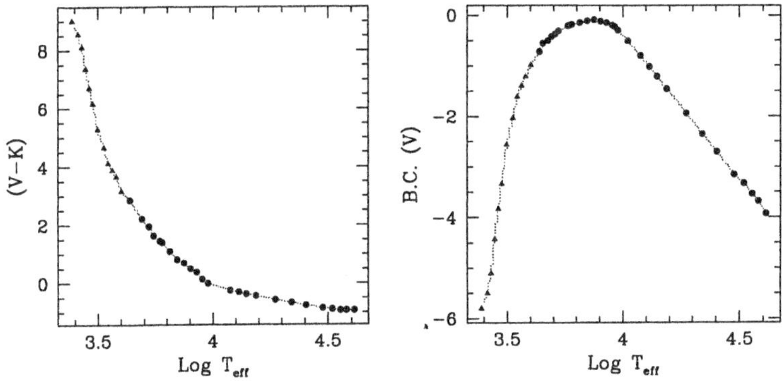

Fig. 24. Adopted BC_V vs. T_{eff} and (V–K) vs. T_{eff} relations, top and bottom panel respectively. Filled triangles are from the M-dwarfs calibration of Bessell ([10]), filled circles are a compilation from the various references given in the text, the dotted line is a linear interpolation fo the data points.

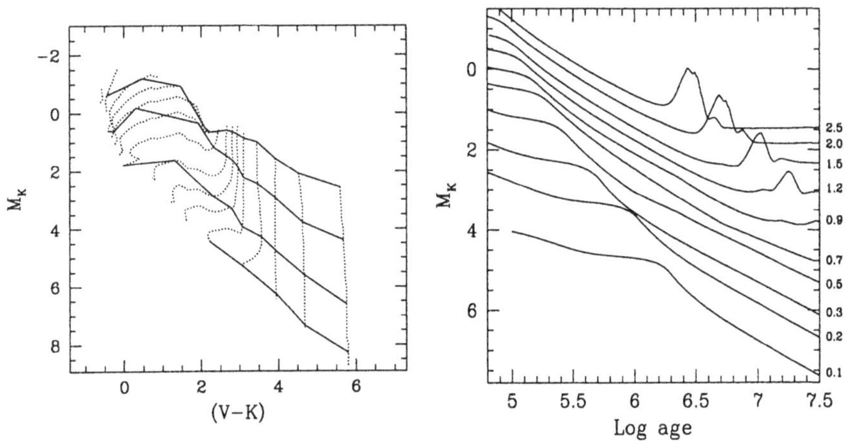

Fig. 25. (*Left panel:*) Colour–magnitude diagram obtained from the evolutionary tracks of Palla & Stahler [42]. The dotted lines show the evolutionary tracks for 0.1, 0.2, 0.4, 0.6, 0.8, 1.0, 1.2, 1.5, 2.0, and 2.5 $M_\odot$, respectively. Solid lines show from top to bottom isochrones for 0.1, 1, 10, and 100 Myr. (*Right panel:*) Absolute K magnitude as a function of time for selected masses, as labeled. (From Testi et al. ([84])

given a K-absolute completeness magnitude, the minimum mass detectable is a function of time: as the age of the cluster increases, one looses sensitivity on the least massive objects. In graphical form this is presented in Fig. 26, where the masses corresponding to M_K are plotted for isochrones between 0.1 and 10 Myr.

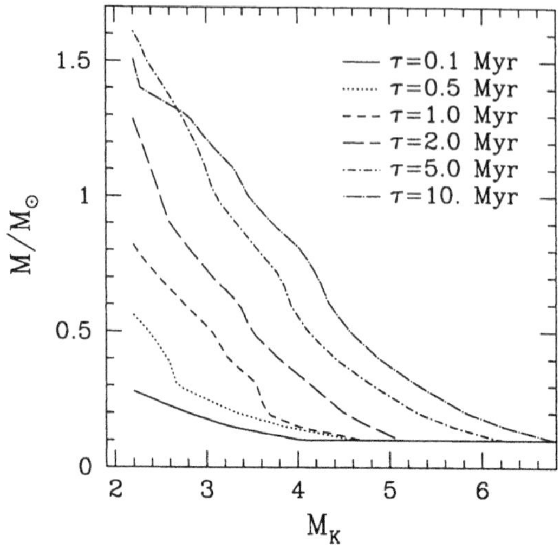

Fig. 26. Stellar mass versus the absolute K magnitude as a function of time. The isochrones in the interval 0.1–10 Myr are indicated. Given the K absolute magnitude, M_K, and the age of a star, it is possible to derive the corresponding limiting mass using the appropriate isochrone.

6.4 Pre-Main-Sequence Binaries

The study of PMS binaries offers the best way to test the theoretical evolutionary tracks. The measurement of the dynamical mass of the components of any binary system yields in fact an absolute mass calibration of the tracks, and thus provides a direct answer for the exact location of the tracks in the H-R diagram. Under the assumption of coeval formation, the derived ages of the binary components test relative age calibrations of the tracks. The population of binary systems in the nearest star forming regions has now been studied and a binary frequency of at least 50% has been derived for systems with angular separations in the range 3 to 1400 AU. It appears that there is even an excess of such systems in comparison to what found among nearby solar-type stars, possibly implying that binary formation is the main mode of star formation.

Mass Estimates Unfortunately, there are very few cases where it has been possible to derive the dynamical mass of the components of an eclipsing binary system. One such case is TY CrA, a binary with a period of 2.89 days in the Corona Australis complex (Beust [13]; Casey [21]). Accurate spectroscopic measurements allow to derive the masses of the individual components, M_1=3.16 $M_\odot$ and M_2=1.64 $M_\odot$. The position of TY CrA in the H-R diagram is shown in Fig. 27. The primary appears slightly below the ZAMS, near a mass of 3.0 $M_\odot$. This value compares well with the more precise spectroscopic determination of 3.16 M$_\odot$. The secondary indeed falls just above the 1.5 M$_\odot$ evolutionary track. Thus, the two methods (spectroscopic and photometric) for obtaining the component masses are in essential agreement. To illustrate the effect of adopting a different set of PMS tracks, Fig. 27 also displays the results of Casey et al. [21] who tested three sets of tracks (Swenson et al. [83], shown in the figure; D'Antona & Mazzitelli [31]; Claret [26]). In all three cases, the tracks for M$_\odot$ =1.64 M$_\odot$ are consistent with the physical parameters. Thus, the secondary star represents the first quantitative dynamical test of PMS tracks. Unfortunately, the accuracies of the derived parameters are not adequate to distinguish among the different sets of tracks.

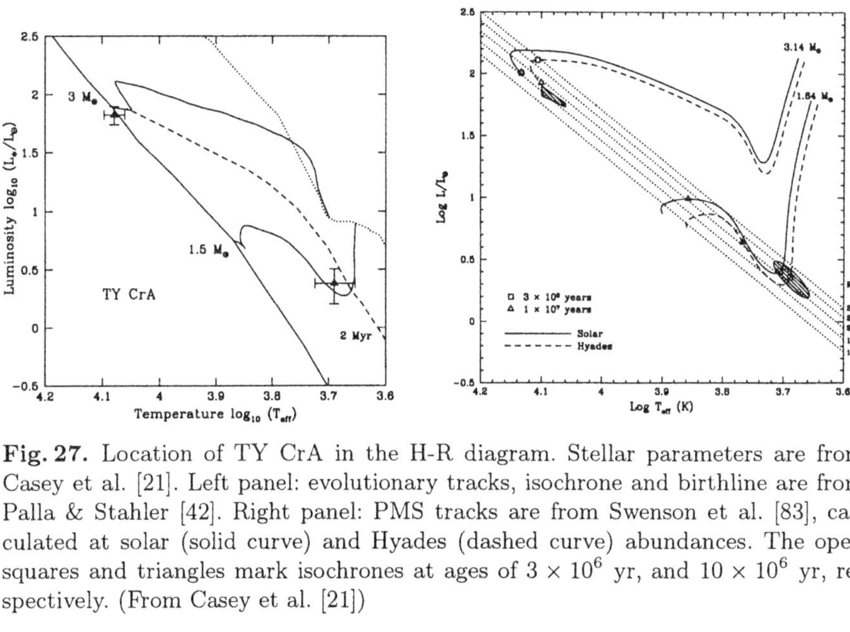

Fig. 27. Location of TY CrA in the H-R diagram. Stellar parameters are from Casey et al. [21]. Left panel: evolutionary tracks, isochrone and birthline are from Palla & Stahler [42]. Right panel: PMS tracks are from Swenson et al. [83], calculated at solar (solid curve) and Hyades (dashed curve) abundances. The open squares and triangles mark isochrones at ages of 3×10^6 yr, and 10×10^6 yr, respectively. (From Casey et al. [21])

A summary of the stellar parameters derived for the best studied cases of PMS binaries is given in Table 2 (Palla & Stahler [70]). Here, we consider both double lined spectroscopic systems for which the highly accurate mass ratios can be derived from the individual radial velocity curves, and eclipsing binaries which yield individual masses. For comparison, the theoretically derived values of masses and mass ratios are given (under the subscript PS). Finally, the last column gives the ages as derived from the isochrones.

Individual masses can also be derived for younger systems which are surrounded by circumstellar disks. The best case is GG Tau, a hierarchical quadruple consisting of two binary stars (Leinert et al. [58]). The close pair, GG Tau A, with a separation 0."25 ($\sim$35 AU), lies 10."1 to the north of a wider pair (separation 1."48$\sim$207 AU). The interest of this system for

Table 2. Properties of spectroscopic binaries

System	T_{eff} (K)	L ($L_\odot$)	M_{dyn} ($M_\odot$)	M_{PS} ($M_\odot$)	$(M_A/M_B)_{dyn}$ ($M_\odot$)	$(M_A/M_B)_{PS}$ ($M_\odot$)	t_{PS} (10^6 yr)
TY CrA	$d = 130$ *pc*						
A	12000±500	67±12	3.16±0.02	2.91	1.93±0.02	$1.82^{+0.11}_{-0.04}$	–
B	4900±400	2.4±0.8	1.64±0.01	1.60			$3.9^{+3.6}_{-2.4}$
EK Cep	$d = 150$ *pc*						
A	9000±200	14.8±1.4	2.03±0.01	1.97	1.82±0.02	$1.73^{+0.15}_{-0.05}$	20
B	5700±200	1.55±0.3	1.12±0.01	1.14			20±4
RS Cha	$d = 97$ *pc*						
A	7810±180	15.2±1.2	1.858±0.016	1.88	1.02±0.02	1.04±0.06	5.0±0.3
B	7295±170	13.9±1.2	1.821±0.018	1.80			$4.3^{+0.8}_{-0.6}$
V773 Tau	$d = 149$ *pc*						
A	4900±180	2.31±0.43		1.53	1.32±0.06	$1.29^{+0.35}_{-0.23}$	4.1±0.9
B	4400±180	1.39±0.43		1.19			$2.8^{+1.6}_{-1.0}$
NTTS 162814	$d = 125$ *pc*						
A	4390	1.20		1.26	1.09±0.07	1.11±0.07	$5.1^{+0.2}_{-2.1}$
B	4083	0.85		1.08			$5.3^{+2.5}_{-0.1}$
P 1540	$d = 470$ *pc*						
A	5010	11.5		1.71	1.32±0.03	1.19±0.18	<0.1
B	4680	7.9		1.25			0.2±0.1
V826 Tau	$d = 140$ *pc*						
A	4100±200	0.55±0.20		0.83	1.02±0.01	1	5.1
B	4100±200	0.55±0.20		0.83			5.1
GG Tau	$d = 140$*pc*						
Aa	4000±150	0.84±0.13	1.28±0.07	0.78			2.3±0.3
Ab	3760±120	0.71±0.09	(total)	0.54			$1.5^{+0.2}_{-0.5}$
Ba	2985±180	0.08±0.02		<0.1			
Bb	2690±190	0.015±0.004		$\ll 0.1$			

testing evolutionary models comes from the dynamical mass estimate for GG Tau A from kinematic studies of its circumbinary disk (Dutrey et al. [37]). Millimeter interferometric data have resolved the circumbinary disk and have yielded a mass of M_{tot} =1.28±0.08 $M_{\odot}$ (D/140 pc) for the mass of the binary stars (Guilloteau et al. [44]).

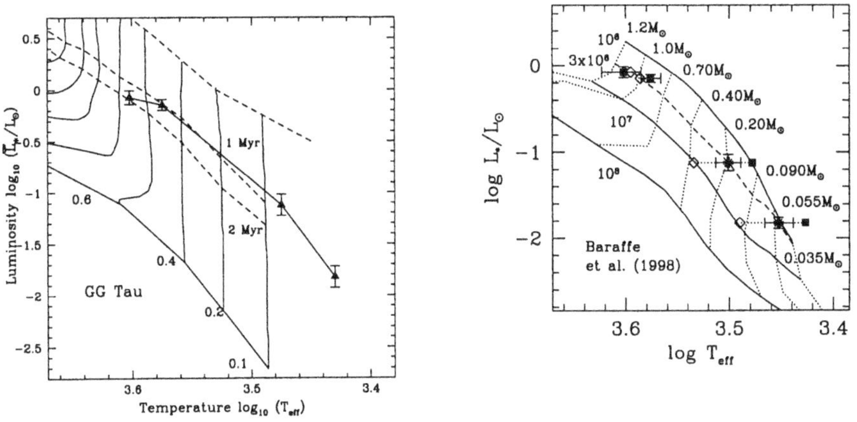

Fig. 28. Location of GG Tau in the H-R diagram. Left panel: comparison with the evolutionary tracks of Palla & Stahler [42], computed with mixing length parameter α =1.5. Right panel: evolutionary tracks of Baraffe et al. [6], with α =1.9. The lowest mass companion has a mass of 0.044±0.04 $M_{\odot}$, assuming coevality for the temperature scale. (From White et al. [87])

In addition, the stellar properties of each of the four components have been determined spectroscopically, using both the HST and Keck telescopes (White et al. [87]). These results confirm that all stars are indeed T Tauri stars, with spectral types ranging from K7 for GG Tau Aa to M7 for GG Tau Bb. Using various temperature scales, White et al. derive the effective temperatures for all stars, while the luminosities have been estimated from the reddening corrected I-band measurements and a bolometric correction, assuming a distance of 140 pc for Taurus-Auriga. The location of the four stars in the H-R diagram is shown in Fig. 28, together with a comparison with the evolutionary tracks of Baraffe et al. [6] (α =1.9). These models yield the most consistent ages, using a temperature scale intermediate between that of dwarfs and giants. The resulting age is 1.5×10^6 yr. The total mass for GG Tau A is 1.46±0.10 $M_{\odot}$, close to the dynamical value obtained by Guilloteau et al. [44]. An estimate of the mass of GG Tau B can then be obtained by assuming that the stars lie on the same isochrone. In this way, the mass of GG Tau Bb results of 0.05 $M_{\odot}$, well below the hydrogen burning minimum mass of ~0.075 $M_{\odot}$. Thus, GG Tau Bb is the coldest, lowest mass, spectroscopically confirmed companion to a T Tauri star.

In conclusion, the analysis shows that PMS tracks are reliable, in the sense that they generally predict masses or mass ratios in agreement with the empirical values.

Age Estimates Derivation of stellar ages comes from theoretical isochrones. here, the discrepancy between different evolutionary tracks are reduced, particularly for systems with ages greater than several million years. For younger stars, the impact of the birthline to set the $t = 0$ mark is largest and can cause differences in derived ages up to an order of a magnitude.

What about the ages of the TY CrA components? Unfortunately, the location of the primary so close to the ZAMS makes it impossible to assign a reliable contraction age. However, the star illuminates a bright reflection nebula in the Corona Australis cloud, and the system's spectral energy distribution rises in the mid-infrared regime toward longer wavelengths. Thus, the primary is undeniably young, a typical Herbig Be star situated within both interstellar and circumstellar gas and dust. Suppose we tentatively assign it the pre-main-sequence age for a 3 $M_\odot$ star just arriving on the ZAMS, or 2×10^6 yr. Then, the corresponding isochrone (dashed curve in Fig. 27) passes close to the secondary's position. The components of TY CrA are therefore plausibly coeval, although the data do not permit a more quantitative assessment.

Age assignments are more trustworthy when both stars lie well above the ZAMS. No eclipsing binaries of this kind are known, but there are a number of non-eclipsing, double-lined systems. In the absence of eclipses, one must obtain $T_{\rm eff}$ and $L_{\rm bol}$ for each star by careful analysis of both the composite SED and the shifting, narrow-band spectrum.

Discerning the relative luminosities from an unresolved pair of stars is no simple matter, and the results are subject to some uncertainty. The situation further improves when we consider wider, spatially resolved binaries. In this case, however, there is no spectroscopic mass ratios for comparison. Studies conducted in several nearby star forming regions have shown that most of the pairs when placed on the H-R diagram appear to be coeval, within the observational errors (Hartigan et al. [46]; Brandner & Zinnecker [16]). This implies that binary systems consist of stars that were born nearly simultaneously. Thus, the typical binary must arise *in situ*, rather than through capture.

The other important effect to consider is the incorrect assignment of the star luminosity that affects the position in the H-R diagram and the age estimate. Since in the past PMS stars were not known to have companions, their luminosity has been overestimated. For the youngest stars on the vertical portion of the Hayashi track the higher luminosity corresponds to a younger age. However, to correct for this effect is not an easy task since binary surveys are usually made photometrically and therefore provide only monochromatic magnitudes (in particular, K-magnitudes). Thus, in order to derive an estimate of the luminosity of the secondary, one has to extrapolate to longer wavelengths which may result in large errors since young stars have large infrared excesses. Within an entire stellar cluster, the error could materially impact the derived star formation history.

By considering only the correction for the K-flux ratios, Simon et al. [79] have derived the contribution of the secondary to the luminosity in a large sample of binary stars in Taurus. They find that the average age of these stars is 2-3 times older than previously thought and that the apparently different distribution in the H-R diagram of singles and binaries is only an artifact of the luminosity overestimate. However, a better treatment of the binarity issue shows that the expected age correction, for a primary of solar mass, is about 1.5, substantially less than the expected value. A useful derivation of the correction factor to the luminosity and age of the primary is given below (see Palla & Stahler [42]).

Statistical Correction of Mass and Age for Unresolved Binaries
Suppose a pre-main-sequence star is observed to have luminosity L_* and effective temperature $T_{\rm eff}$. Suppose also that this object is known to be on the convective (vertical) portion of its evolutionary track. From the temperature and luminosity, one can estimate the star's age given by the Kelvin-Helmholtz time. For convective, solar-type PMS stars, the effective temperature is related to the mass by the approximate relation:

$$T_{\rm eff} = \bar{T}\left(\frac{M_*}{\bar{M}}\right)^n , \qquad (46)$$

where $n \sim 0.2$, and where $\bar{M}$ and $\bar{T}$ are a fiducial mass and temperature, respectively. To relate R_* to $T_{\rm eff}$ and L_*, we may use the blackbody law and we find that the age, t, of the star is given by:

$$t = \frac{G\bar{M}^2\,(T_{\rm eff}/\bar{T})^{2/n}\,(4\pi\sigma T_{\rm eff}^4)^{1/2}}{L_*^{3/2}} , \qquad (47)$$

Now, it turns out that the star is not a single object, but an unresolved binary. Let q be the ratio of the mass of the unseen companion, M_2, to the mass of the primary star, M_1. What is the actual luminosity of the two stars as a function of q? Also, given a reasonable distribution of secondary masses, what is the *average* luminosity ratio?

To answer the first question, assume that the two stars have precisely the same age. If T_2 is the temperature of the unseen companion, then, since the two luminosities, L_1 and L_2, together add to give L_*, the equal-age requirement becomes, by eq. (47),

$$\frac{(T_1/\bar{T})^{2/n}\,T_1^2}{L_1^{3/2}} = \frac{(T_2/\bar{T})^{2/n}\,T_2^2}{(L_* - L_1)^{3/2}} . \qquad (48)$$

Using eq. (46) to evaluate the temperatures, the expression for L_1 is:

$$\frac{L_1}{L_*} = \left[1 + q^{4(1+n)/3}\right]^{-1} \equiv f(q) \ . \tag{49}$$

In order to compute the reduction factor $f(q)$, the distribution of secondary masses must be specified. One can adopt the initial mass function for field stars, an assumption that apparently holds for at least G-type and M-type primaries (Duquennoy & Mayor [36]; Fischer & Marcy [38]). The average value of $f(q)$, is then

$$f \equiv \int_0^1 f(q)\, \xi(q)\, dq \ . \tag{50}$$

where

$$\xi(q) \equiv \xi(qM_1/M_\odot)\, M_1/M_\odot \ , \tag{51}$$

so that $\int_0^\infty \xi(q)\, dq = 1$. Note that f depends on the primary mass M_1, and may readily be evaluated numerically from eq. (50). The factor is only small for the lowest mass-values, and rises to 0.75 by $M_1 = 0.08\, M_\odot$. Thus, the primary luminosity could be overestimated by at most 25 percent.

One may also estimate the error in age introduced by an unseen companion. From eq. (47), we first obtain $t_1(q)$, the true primary age as a function of q:

$$t_1(q) = \frac{G\bar{M}^2\, (T_{\rm eff}/\bar{T})^{2/n}\, (4\pi\sigma T_{\rm eff}^4)^{1/2}}{[f(q)]^{3/2}\, L_*^{3/2}} \ . \tag{52}$$

Assume that the observed surface temperature corresponds to that of the primary, i.e. $T_1 = T_{\rm eff}$, The ratio of the true age to the naïve value $t_\circ$ is then

$$\frac{t_1(q)}{t_\circ} = [f(q)]^{-3/2} \ , \tag{53}$$

where one has made use of the definition of $t_{\rm KH}$. Now, the *maximum* age discrepancy corresponds to $q = 1$. Evaluating $f(1)$ from eq. (49), the corresponding ratio t_1/t_0 is 2.83. An age ratio of 2.0 requires a q-value of 0.7. Such relatively high companion masses are however unlikely.

The average age discrepancy is obtained by folding in the expected distribution of mass ratios, $\xi(q)$. Let us consider the case of a primary mass near $1\ M_\odot$. According to Duquennoy & Mayor [36], the appropriate q-distribution is approximately Gaussian:

$$\xi(q) = H \exp\left(-\frac{(q-\mu)^2}{2\sigma^2}\right) \ , \tag{54}$$

where $\mu = 0.23$, $\sigma = 0.42$, and H is the normalization constant. The mean value of q, according to eq. (54), is only 0.23. The average age ratio is therefore:

$$\langle \frac{t_1}{t_\circ} \rangle = \int_0^1 [f(q)]^{-3/2}\, \xi(q)\, dq \, . \tag{55}$$

The integral can be evaluated numerically, and the resulting ratio is 1.46. In conclusion, for realistic mass distributions of the secondary, the correction to the mass does not exceed 10–20%, while ages can be off by only a factor ~1.5.

References

1. Alexander, D.R., Ferguson, J.W. 1994, ApJ, 437, 879
2. Alexander, D.R., Augason, G.C., Johnson, H.R. 1989, ApJ, 345, 1014
3. Alexander, D.R., Tamanai, A.I., Allard, F., Ferguson, J.W. 1998, BAAS, 67, 16
4. Allard, F., Hauschildt, P.H., Alexander, D.R., Starrfield, S. 1997, ARAA, 35, 137
5. Baraffe, I., Chabrier, G., Allard, F., Hauschildt, P.H. 1997, A&A, 327, 1054
6. Baraffe, I., Chabrier, G., Allard, F., Hauschildt, P.H. 1998, A&A, 337, 403
7. Beech, M., Mitalas, R. 1995, ApJS, 95, 517
8. Bernasconi, P., Maeder, A. 1996, A&A, 307, 829
9. Bessell, M.S. 1979, PASP, 91, 589
10. Bessell, M.S. 1991, PASP, 101, 66
11. Bessell, M.S. 1998, in Fundamental Stellar Properties, IAU Symp. 189, eds. T.R. Bedding, A.J. Booth & J. Davis (Dordrecht: Kluwer Acad. Press), p. 127
12. Bessell, M.S., Brett, J.M. 1988, PASP, 100, 1134
13. Beust, H., Corporon, P., Siess, L., Forestini, M., Lagrange, A.-M. 1997, A&A, 320, 478
14. Böhm-Vitense, E. 1958, Zs.f.Ap., 46, 108
15. Bonnell, I., Bate, M.R., Zinnecker, H. 1998, MNRAS, 298, 93
16. Brandner, W., Zinnecker, H. 1997, A&A, 321, 220
17. Briceno, C., Hartmann, L., Stauffer, J., Martín, E. 1998, AJ, 115, 2074
18. Burrows, A., Hubbard, W.B., Saumon, D., Lunine, J.I. 1996, ApJ, 406, 158
19. Canuto, V.M., Mazzitelli, I. 1991, ApJ, 370, 295
20. Canuto, V.M., Goldman, I., Mazzitelli, I. 1996, ApJ, 473, 550
21. Casey, B.W., Mathieu, R.D., Vaz, L.P.R., Andersen J., Suntzeff, N.B. 1998, AJ, 115, 1617
22. Caughlan, G., Fowler, W. 1988, Atomic Data Nucl. Data Tables, 40, 283
23. Chabrier, G., Baraffe, I. 1995, ApJ, 451, L29
24. Chabrier, G., Baraffe, I. 1997, A&A, 327, 1039
25. Chabrier, G., Baraffe, I., Plez, B. 1996, ApJ, 459, L91
26. Claret, A. 1995, A&A, 109, 441
27. Cohen, M., Kuhi, L.V. 1979, ApJS, 41, 743
28. Cox, A.N., Tabor J.E. 1976, ApJS, 31, 271
29. D'Antona, F. 1993, in Inside the Stars, eds. W.W. Weiss & A. Baglin, A.S.P. Conf. Ser. 40 (San Francisco: Astr. Soc. Pac.), p. 395
30. D'Antona, F., Mazzitelli, I. 1984, A&A, 138, 431
31. D'Antona, F., Mazzitelli, I. 1994, ApJS, 90, 467
32. D'Antona, F., Mazzitelli, I. 1998, in Cool Stars in Clusters and Associations, eds. G. Micela, R. Pallavicini & S. Sciortino, Mem.S.A.It., 68, 807
33. Däppen, W. 1993, in Inside the Stars, eds. W.W. Weiss & A. Baglin, A.S.P. Conf. Ser. 40 (San Francisco: Astr. Soc. Pac.), p. 208

34. Delfosse, X., Forveille, T., Mayor, M., Burnet, M., Perrier, C. 1999, A&A, 341, L63
35. Draine, B.T., Lee, H.M. 1984, ApJ, 285, 89
36. Duquennoy, A., Mayor, M. 1991, A&A, 248, 485
37. Dutrey, A., Guilloteau, S., Simon, M. 1994, A&A, 286, 149
38. Fischer, D. A., Marcy, G. W. 1992, ApJ, 396, 178
39. Forestini, M. 1994, A&A, 285, 473
40. Gizis, J.E. 1997, AJ, 113, 806
41. Graboske, H., DeWitt, H.E., Grossman, H.S., Cooper, M.S. 1973, ApJ, 181, 457
42. Grevesse, N., Sauval, A.J. 1998, SSRv, 85, 161
43. Grossman, A.N., Graboske, H.C. 1971, ApJ, 164, 475
44. Guilloteau, S., Dutrey, A., Simon, M. 1999, A&A, 348, 570
45. Habets, G.M.H.J., Heintze, J.R.W. 1981, A&AS, 46, 193
46. Hartigan, P., Strom, K.M., Strom, S.E. 1994, ApJ, 427, 961
47. Henry, T.J., McCarthy, D.W., Jr. 1993, AJ, 106, 773
48. Huebner, W.F., Merts, A., Magee, N.H., Argo, M.F. 1977, Los Alamos Sci. Lab. Rep., LA-6760-M
49. Iglesias, C.A., Rogers, F.J. 1991, ApJ, 371, L73
50. Iglesias, C.A., Rogers, F.J. 1996, ApJ, 464, 943
51. Kahn, F.D. 1974, A&A, 37, 149
52. Kenyon, S.J., Hartmann, L. 1995, ApJS, 101, 117
53. Kirkpatrick, J.D., Kelly, D.J., Rieke, G.H., Liebert, J. 1993, ApJ, 402, 643
54. Koorneeff, J. 1983, A&A, 128, 84
55. Lacy, C.H. 1977, ApJ, 218, 444
56. Larson, R.B., Starrfield, S. 1971, A&A, 13, 190
57. Leggett, S.K., Allard, F., Berriman, G., Dahn, C.C., Hauschildt, P. 1996, ApJS, 104, 117
58. Leinert, C., Zinnecker, H., Weitzel, N., Christou, J., Ridgway, S.T., Jameson, R., Haas, M., Lenzen, R. 1993, A&A, 278, 129
59. Leung, K.-C., Schneider, D.P. 1978, AJ, 83, 618
60. Luhman, K.L., Rieke, G.H. 1998, ApJ, 497, 354
61. Maceroni, C., Rucinski, S.M. 1997, PASP, 109, 782
62. Norberg, P., Maeder, A. 1999, A&A, 359, 1025
63. Behrend, R., Maeder, A. 2000, A&A, 373, 190
64. Mazzitelli, I. 1989, in Low-Mass Star Formation and Pre-Main-Sequence Objects, ed. Bo Reipurth, ESO Proc. 33, p. 407
65. Mazzitelli, I., Moretti, M. 1980, ApJ, 235, 955
66. Metcalfe, T.S., Mathieu, R.D., Latham, D.W., Torres, G. 1996, ApJ, 456, 356
67. Meyer, M.R. 1996, PhD Thesis, University of Massachussets
68. Mihalas, D., Däppen, W., Hummer, D.G. 1988, ApJ, 331, 815
69. Nakano, T. 1989, ApJ, 345, 464
70. Palla, F., Stahler, S.W. 2001, ApJ, 553, 299
71. Pollack, J.B., Hollenbach, D., Beckwith, S., Simonelli, D.P., Roush, T., Fong, W. 1994, ApJ, 421, 615
72. Rogers, F.J., Iglesias, C.A. 1992, ApJS, 79, 507
73. Salpeter, E.E. 1954, Mem. Roy Soc. Liège, 14, 116
74. Schaller, G., Schaerer, D., Meynet, G., Maeder, A. 1992, A&AS, 96, 269
75. Schmidt-Kaler, Th. 1981, in Landolt-Börnstein, Gruppe VI, Bond 2,1

76. Seaton, M.J. 1993, in Inside the Stars, eds. W.W. Weiss & A. Baglin, A.S.P. Conf. Ser. 40 (San Francisco: Astr. Soc. Pac.), p. 222
77. Seaton, M.J., Yan, Y., Mihalas, D., Pradhan, A.K. 1994, MNRAS, 266, 805
78. Simon, N.R. 1982, ApJ, 260, L87
79. Simon, M., Ghez, A.M., Leinert, Ch. 1993, ApJ, 408, L33
80. Stahler, S.W., Palla, F., Ho, P.T.P. 2000, in Protostars and Planets IV, eds. V. Mannings, A.P. Boss, and S.S. Russell (Tucson: U. of Arizona Press), p. 327
81. Stauffer, J., R., Hartmann, L., Barrado y Navàscues, D. 1995, ApJ, 454, 910
82. Strom, K.M., Strom, S.E. 1994, ApJ, 424, 237
83. Swenson, J.H., Faulkner, J., Rogers, F.J., Iglesias, C.A. 1996, priv. comm.
84. Testi L., Palla F., Natta A. 1998, A&AS, 133, 81
85. Ventura, P., Zeppieri, A., Mazzitelli, I., D'Antona, F. 1997, A&A, 331, 1011
86. Viti, S., Jones, H.R.A., Schweitzer, A., Allard, F., Hauschildt, P.H., Tennyson, J. 1997, MNRAS, 291, 780
87. White, R.J., Ghez, A.M., Reid, I.N., Schultz, G. 1999, ApJ, 520, 811
88. Wilking, B.A., Greene, T.P., Meyer, M.R. 1999, AJ, 117, 469
89. Wolfire, M.G., Cassinelli, J.P. 1987, ApJ, 319, 850
90. Yorke, H., Krügel, E. 1977, A&A, 54, 183

III. Physical Processes in PMS Stars

7 Lithium Depletion in Young Stars

The depletion of lithium in both young and evolved stars is considered one of the strongest constraints imposed on stellar evolution theory. Surface abundances in fact provide a critical test for the physical processes (convection, opacity, rotation, mass loss) which ought to be included in the theoretical models in order to reproduce the observed level of depletion. Due to its small binding energy, ^{7}Li (the most abundant of the two stable isotopes of lithium) is easily destroyed in stellar interiors via the reaction ^{7}Li(p,α)^{4}He at temperatures around 2.5×10^6 K (Bodenheimer [8]). According to standard models, depletion occurs only when the outer convection extends down to that temperature. Thus, it appears quite unavoidable that some ^{7}Li depletion due to nuclear burning takes place already during PMS contraction. Other processes, associated with the mixing due to transport of angular momentum through the star, are effective on much longer time scales and dominate the depletion history during the main sequence phase (e.g. Forestini [39]; Martín & Claret [48]; Ventura et al. [85]).

The theoretical predictions are rather well established, independently of the adopted models. Stars more massive than 1.7 $M_\odot$ should *not* show any ^{7}Li depletion at all; solar-type stars are significantly depleted, while for sub-solar stars there should be no ^{7}Li left over. For stars with $M_*\gtrsim0.5$ $M_\odot$, most lithium burns after convection has halted in the stellar core. Predictions of lithium depletion then depend on the temperature at the bottom of the convection zone, whose exact location is sensitive to opacity, convection, and rotation. Stars with $M_*\lesssim0.5$ $M_\odot$ are fully convective during lithium burning. Since the rate of contraction of such stars is controlled by the effective temperature, the amount of lithium depletion is also sensitive to $T_{\rm eff}$ and hence to the surface opacities.

Several groups have followed numerically the phase of lithium burning, and the models make specific predictions about the time evolution of the lithium depletion boundary. At ages 30, 70 and 140 Myr, the lithium depletion edge should occur at 0.17 $M_\odot$, 0.09 $M_\odot$ and 0.07 $M_\odot$ (Baraffe et al. [5]; Burrows et al. [15]; D'Antona & Mazzitelli [32]). Figure 29 illustrates the behavior of a 1 $M_\odot$ star according to the models of D'Antona & Mazzitelli [32] for variations of the input parameters. In all cases, the amount of lithium burning is very large, with the depletion factor ranging between 100 and 400. Thus, most lithium should be consummed during the PMS phase, a result with is at odds with the observational evidence of the surface abundances in open clusters of ages of about 5×10^7–10^8 yr. Also note the rather surprising dependence of the results on the metallicity of the gas mixture: a change of only $\sim$10% can cause an order of magnitude difference in the depletion factor.

To bypass the uncertainties of numerical models, Bildsten et al. [7] have developed a clever analytic approach based on the fact that $T_{\rm eff}$ remains

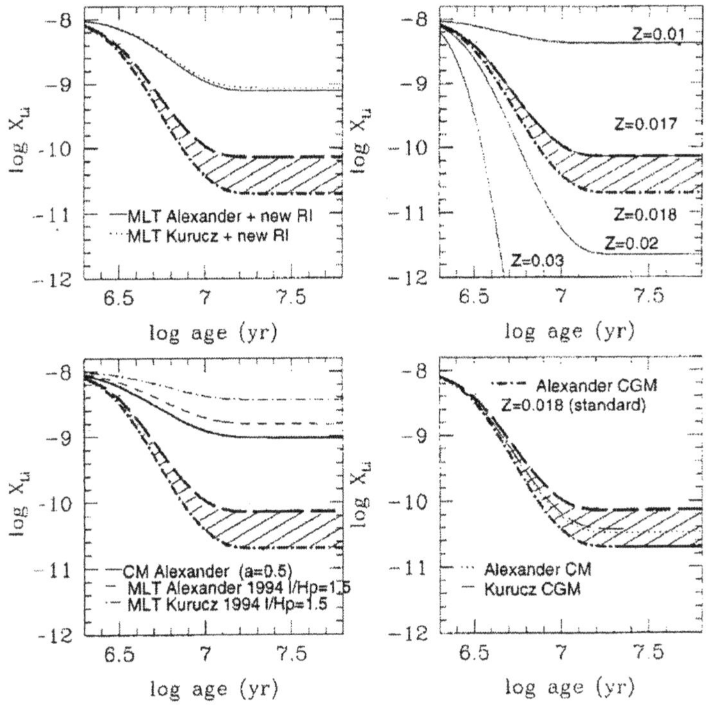

Fig. 29. The dependence of ^{7}Li-depletion as function of time for a 1 $M_\odot$ PMS star. The four panels illustrate the sensitivity on opacity (upper left), metallicity (upper right), prescription for convection (lower left), and a combination of opacity and turbulence. The shaded area in all models represents the standard solar model with Y=0.28, Z=0.017-0.018, and the turbulent spectrum of Canuto et al. (1996). (From D'Antona & Mazzitelli [32])

approximately constant during the fully convective phase (see the low-mass tracks of Fig. 14). Rather than calculating $T_{\rm eff}$, Bildsten et al. calculate the dependence of lithium depletion on $T_{\rm eff}$, obtaining the radius, age and luminosity of a PMS star as a function of lithium depletion. This method can be used to derive the age of open clusters with higher accuracy than any other method.

7.1 Rotation and Depletion

Another important aspect of the lithium problem is related to the rotational properties of PMS stars. Studies of low-mass stars in the Pleiades and Hyades have shown that *rapid* rotators have more lithium than slow rotators (e.g. Barrado y Naváscues & Stauffer [4]). This correlation breaks down at lower masses. The early results of PMS models including angular momentum transfer in the interior and angular momentum loss through stellar winds showed that lithium depletion is *enhanced* with respect to nonrotating models (Pinsonneault et al. [58]). However, the models of Martín & Claret [48], which included solid body rotation and no mass loss, obtained the opposite result whereby lithium depletion is *reduced*. More recently, Mendes et al. [51] have confirmed the early results of Pinsonneault et al., confirming that it is unlikely that the lithium depletion problem can be solved by rotation alone. As a general result, the effect of rotation is to decrease the level of lithium depletion in fully convective interiors, and to increase the efficiency of burning as soon as the star develops a radiative core.

7.2 Spectroscopic Classification Based on Lithium

The equivalent width of the of the photospheric LiI line at 6708 Åcan also be used as a spectroscopic criterion for the classification of low-mass PMS stars (Martín [47]). Such a diagnostic is important since it is not always possible to distinguish young stars in different evolutionary phases. The problem has become even more complicated with the discovery of many X-ray emitting stars possibly associated with PMS stars far from molecular clouds (Neuhäuser [54]) whose nature and origin are still unclear. This problem is closely linked to the puzzle of the missing post-T Tauri stars in associations, i.e. stars with an age greater than $\sim 10^7$ yr (Herbig [31]).

Martín has shown that T Tauri stars occupy different regions in the $T_{\rm eff}$–$W_{\rm LiI}$ diagram, as shown in Fig. 30. Most T Tauri stars fall on or above the lithium isoabundance line at log N(Li)=2.8 (in the scale of log N(H)=12), which represents the minimum abundance observed in a large sample of stars (Magazzú et al. [42]; Martín et al. [49]). For masses $\lesssim$0.5 $M_\odot$, models predict that Li depletion is not significant until ages larger than $\sim 10^7$ yr. Since a typical T Tauri star has a mass of $\sim$0.5 $M_\odot$ and an age much less than this value, it has preserved most of its initial lithium content, which should therefore be easily detectable as a strong absorption line. On the other hand, low–mass members of older system, such as the open clusters displayed in the figure, show lithium equivalent width much weaker than T Tauri stars and the separation between the two groups becomes increasingly larger at lower $T_{\rm eff}$. Finally, post-T Tauri stars can be easily identified if they fill the empty region between the T Tauri stars and the low-mass members of open clusters. Martín has dubbed this part of the diagram the "post-T Tauri gap". This way, accurate $W_{\rm LiI}$ measurements at high spectral resolution can provide the spectral diagnostic to obtain more reliable estimates of the fraction of WTTS and post-T Tauri stars in and around star forming regions.

7.3 Lithium and Brown Dwarfs

One of the most successful applications of the theory of Li-burning has been in the field of brown dwarfs. The limiting mass for ^{7}Li-burning is $M_* \geq$0.06

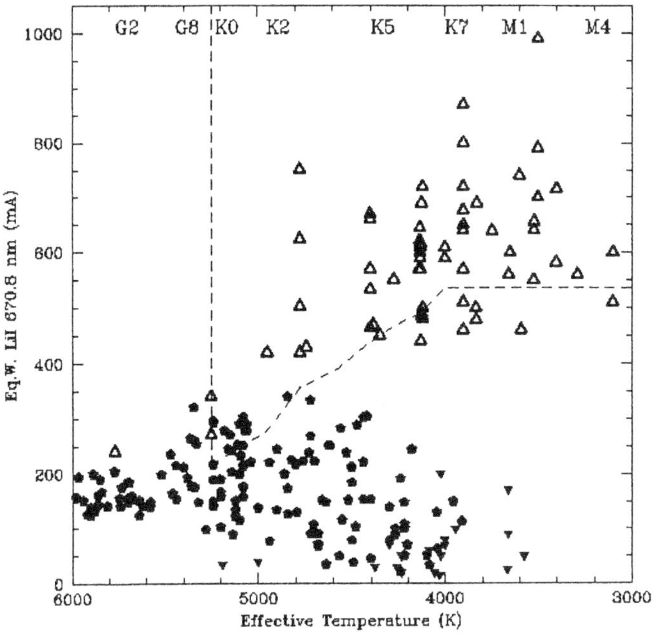

Fig. 30. The equivalent widths of the LiI 6708 Åabsorption line of low-mass stars as a function of T_{eff}. T Tauri stars are shown by the empty triangles and low-mass mass members of various young open clusters by filled symbols. The dashed line represents the locus of minimum equivalent width values for log N(Li)=2.8, and the cutoff for the maximum T_{eff} =5250 K of T Tauri stars. (From Martín [47])

$M_\odot$, since degeneracy effects in lower mass stars prevent the onset of nuclear reactions. This value is below the substellar mass limit, usually taken between 0.08 and 0.07 $M_\odot$. Thus, it has been suggested that the presence of lithium lines in the observed stellar spectra can be used as a powerful probe to search for brown dwarf candidates (Rebolo, Martín & Magazzù [61]). The Li-depletion history for different masses is shown in Fig. 31. Using this diagnostic, the first brown dwarfs have been discovered in the Pleiades (Rebolo et al. [63]; Basri, Marcy & Graham [6]) and about a dozen candidates have been identified to date (Martín et al. [50]; Stauffer et al. [73]). The object PPl 15 in the Pleiades is particularly interesting: it is the first confirmed brown dwarf by the lithium test and the first brown dwarf double lined spectroscopic binary (Basri & Martín [5]). The derived period is about 5.8 days and the eccentricity ~0.4. The dynamical mass ratio is about 0.85, yielding individual masses between 60 and 70 Jupiter masses. The very short period of the system poses new challenges to binary formation mechanisms.

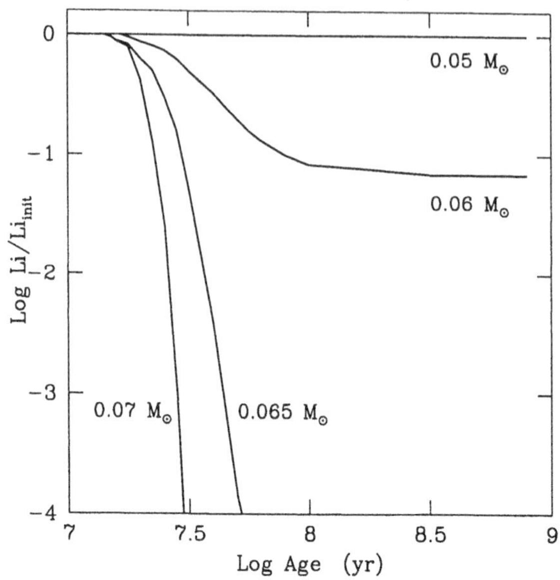

Fig. 31. ^{7}Li-depletion as function of time for stars in the mass range 0.07 to 0.05 $M_\odot$. (From Rebolo et al. [63])

8 Mass Accretion in Young Stars

The evolution described so far, although improved by the account of more realistic initial conditions and input physics, is still rather conventional. It assumes that each star evolves in isolation at constant mass from the birthline to the arrival on the main sequence. But we know that in reality the situation is more complex, owing to the interaction of the star with the environment. In particular, we have neglected both the thermal effect of stellar winds and their associated mass loss. Empirical determination of the mass outflow in T Tauri stars remains problematic, with most estimates utilizing the high velocity component of the optical forbidden line emission occurring within 100 AU of the star. The studies of Hartigan, Edwards & Ghandour [27] and Hirth, Mundt & Solf [34] yield rates of order $\dot{M}_{\rm loss} \sim 10^{-9}$ $M_\odot$ yr^{-1} for a sample of about 40 T Tauri stars. Since the typical age of these stars is a few $\times 10^6$ yr, their diminution of mass has been relatively small.

8.1 Disk Accretion

Another equally important effect to consider is the occurrence of residual mass accretion from a circumstellar disk during the PMS phase. Accretion alters the total (star+disk) luminosity, the evolutionary path, and hence the age estimate of a star. Hartmann, Cassen & Kenyon [19] have presented a detailed analysis of the effects of disk accretion on the initial evolution of fully convective low-mass PMS stars.

Current estimates of mass accretion rates derive from the analysis of the excess optical continuum radiation in T Tauri and Herbig stars that also show an infrared excess. The latter is interpreted as being due to the presence of an active circumstellar disk. In this case, the disk luminosity is the sum of two contributions,

$$L_{\rm disk} = \frac{GM_*\dot{M}_{\rm acc}}{2R_*} + L_{\rm rep}, \tag{56}$$

where the first term represents the accretion luminosity, while the second term is due to the absorption and reradiation of light from the central star. Typically, $L_{\rm rep}$ ~0.25–0.5 L_*. Thus, unless $L_{\rm disk} \geq L_*$, the identification of the excess luminosity as *accretion* is not secure. On the other hand, observations indicate that at most $L_{\rm disk}$ ~0.5 L_*.

If the optical excess is assumed to arise in a boundary layer (BL) between the star and a disk, then the standard theory of viscous accretion disks predicts that up to half of the accretion luminosity is emitted as radiation,

$$L_{\rm BL} \leq \frac{GM_*\dot{M}_{\rm acc}}{2R_*}. \tag{57}$$

Thus, measurements of $L_{\rm BL}$ can provide a direct estimate of the amount of mass actually falling onto a star. In practice, estimates of the accretion rates come from measurements of the hot continuum excess produced as disk material impacts on the stellar surface. This method requires a careful separation of the contribution due to accretion produced emission from that intrinsic to the stellar photosphere. This is a difficult procedure when the accretion luminosity is small compared to the stellar luminosity, which is always the case even in the youngest stars. In addition, both extinction corrections and assumptions on the geometry of accretion may lead to estimates of the accretion rates that are significantly uncertain, and that may vary from author to author up to an order of magnitude for the same object. Recently, Gullbring et al. [26] and Hartmann et al. [29] have presented a careful analysis of ultraviolet, optical and infrared lines in a sample of 60 stars in Taurus and Chamaeleon cloud complexes. The derived mass accretion rates for 40 objects are shown in Fig. 32. The typical accretion rate is $\dot{M}_{\rm acc}$ ~10^{-8} $M_\odot$ yr^{-1}, again too small for significant mass change during the earliest, active phases of contraction.

Higher values of $\dot{M}_{\rm acc}$, up to 100 times, have been derived for the more massive Herbig Ae/Be stars (e.g. Hillenbrand et al. [33]). These estimates are based on infrared disk emission, assuming that the excess emission is mostly due to viscous dissipation. However, it is often the case that irradiation from the central star provides more energy than internal dissipation, so that the accretion rates tend to be overestimated. Note also that, in the absence of angular momentum loss, accretion at large rates would spin up the central

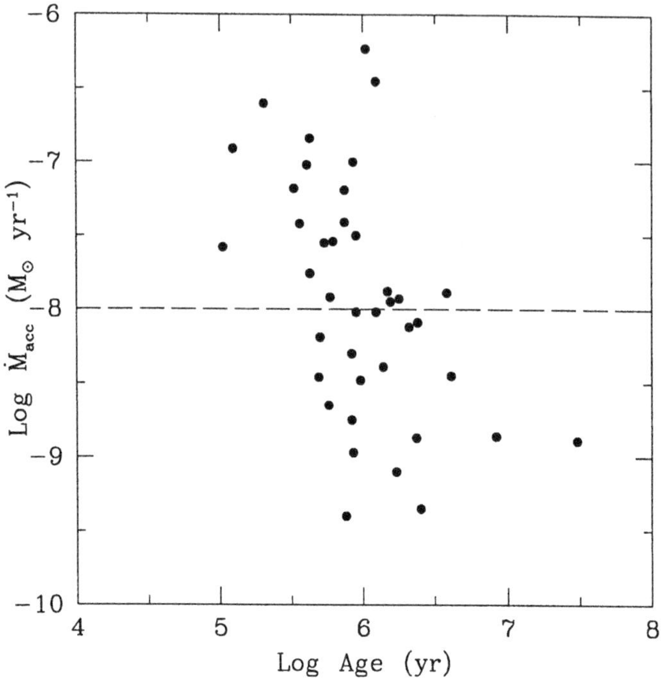

Fig. 32. Mass accretion rates for T Tauri stars. Values of $\dot{M}_{\rm acc}$ for 40 stars are plotted as a function of stellar age. The median accretion rate for T Tauri stars of age $\sim 10^6$ yr is $\sim 10^{-8}$ $M_\odot$ yr^{-1}. (From Hartmann et al. [29])

star to half the break-up velocity in a few million years. Since young stars are known to rotate at typically 1/10–1/5 of the break-up speed, it is likely that strong winds carry away the angular momentum excess. Therefore, it is still not clear what fraction of the accreted mass is actually incorporated into the star.

8.2 Effects on Evolutionary Tracks

The question, then, is whether the inferred rates are high enough to modify the evolutionary tracks. Hartmann & Kenyon [28] argue that indeed the path is greatly affected, and estimate from qualitative arguments that about 10% of the final mass of a T Tauri is accreted during the PMS phase. Simple energy considerations allow to derive an estimate of the effect of mass accretion on fully convective stars (e.g. Palla [55]; Hartmann, Cassen & Kenyon [19]). The evolution of the stellar luminosity is described by the equation

$$\frac{d}{dt}\left(-\frac{3GM_*^2}{7R_*}\right) = -L_* - \frac{GM_*\dot{M}_{\rm acc}}{R_*} + \dot{M}_{\rm acc}\delta\ , \qquad (58)$$

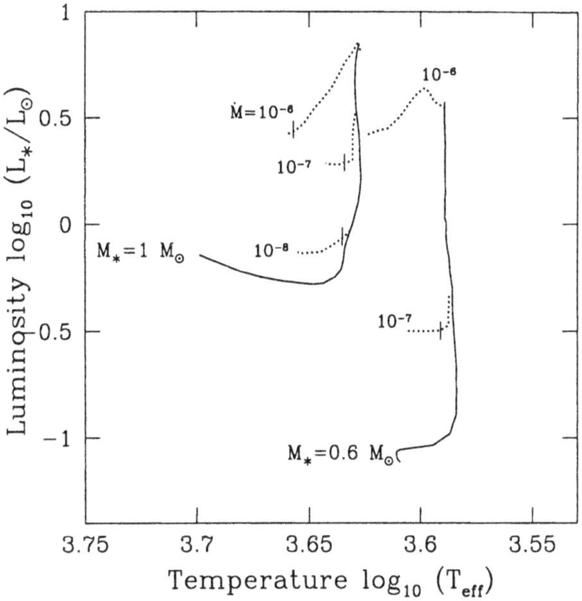

Fig. 33. Effects of mass accretion during PMS evolution. The solid curves represent the standard tracks at constant mass for M_* =0.6 and 1.0 $M_\odot$. Three values of $\dot{M}_{\rm acc}$ are considered, 10^{-8}, 10^{-7}, and 10^{-6} $M_\odot$ yr^{-1}. The resulting tracks are shown by the dashed lines. The tickmarks correspond to an age of 3×10^6 yr. (From Parigi [56])

where the term on the left hand side represents the temporal change of the gravitational potential energy; the second and third terms on the rhs express the gain of (negative) energy by mass accretion and the released luminosity of freshly accreted deuterium. The main difference with the heat equation (36) is now in the interpretation of the time derivative of the entropy which can no longer be computed at constant mass. Because of the addition of new mass elements, it is more convenient to introduce a *relative mass* coordinate $q \equiv m/M$, such that

$$\left(\frac{\partial s}{\partial t}\right)_m = \left(\frac{\partial s}{\partial t}\right)_q - q\,\dot{M}_{\rm acc}\left(\frac{\partial s}{\partial m}\right)_t . \tag{59}$$

From the definition of q

$$\left(\frac{\partial m}{\partial t}\right)_q = \dot{M}q\,, \tag{60}$$

and the heat equation becomes

$$\left(\frac{\partial s}{\partial m}\right)_t = \epsilon - T\left(\frac{\partial s}{\partial t}\right)_q + q\,\dot{M}_{\rm acc}\,T\left(\frac{\partial s}{\partial m}\right)_t . \tag{61}$$

The outer boundary conditions can be either those of a normal photosphere as in eqs. (41) and (42), or those appropriate to an accretion shock. A condition is that the pressure at the stellar surface must equal the ram pressure due to infalling matter, given by $\rho V_{\rm ff}^2$, where $V_{\rm ff}$ is the free-fall velocity. Then,

$$P(M_*) = \frac{\dot{M}_{\rm acc}}{4\pi}\left(\frac{2GM_*}{R_*^5}\right)^{1/2} . \tag{62}$$

The last boundary condition concerns the surface value of the temperature and its relation to the luminosity. The total luminosity of an accreting PMS star is given by two contributions

$$L_* = L_{\rm post} + L_{\rm acc} , \tag{63}$$

where $L_{\rm acc}$ is the luminosity released by accretion, and $L_{\rm post}$ is that radiated by the interior

$$L_{\rm post} = 4\pi R_*^2 \sigma_B T_{\rm post}^4 - \frac{3}{4} L_{\rm acc} , \tag{64}$$

and $T_{\rm post}$ is the postshock temperature.

In order to test the effect of mass accretion, the evolution of stars with initial mass of 0.6 and 1.0 $M_\odot$ has been followed allowing for steady mass accretion at different rates for the first 3 million years (Parigi [56]). The net result of fast ($\dot{M} \sim 10^{-6}$ $M_\odot$ yr^{-1}) accretion during contraction is to accelerate the evolution of a star, i.e. a star reaches a given evolutionary point at an age smaller than that of a star evolving at constant mass. Conversely, for low accretion rates ($\dot{M} \sim 10^{-8}$ $M_\odot$ yr^{-1}) a star experiences a delay in reaching the same evolutionary state, and therefore appears relatively older. The resulting evolution is displayed in Fig. 33. As we can see, as long as the value of $\dot{M}_{\rm acc}$ remains below $\sim 10^{-7}$ $M_\odot$ yr^{-1}, the new tracks do not differ substantially from those computed at constant mass. Moreover, if the *global* effect of mass accretion is limited to about 10% of the final mass, it would take only 1 million years to a 1 $M_\odot$ star to accrete such an amount of mass at $\dot{M}_{\rm acc}=10^{-7}$ $M_\odot$ yr^{-1}. Small departures can be appreciated only after at least 2 million years for $\dot{M}_{\rm acc}=10^{-7}$ $M_\odot$ yr^{-1}, implying that, unless accretion continues for a substantial part of the PMS evolution at high rates, the uncertainty in the location of a star in the H-R diagram due to this effect is small.

Following a different approach, Siess & Forestini [66] and Siess, Forestini & Bertout [68] have computed the evolution of stars that accrete a substantial part of their final mass during the PMS phase. In such case the evolutionary tracks are greatly affected: compared to a non-accreting star, the evolution is accelerated, the central temperature remains larger, and the star appears systematically younger (with age estimates affected by up to factors 2 to 3). The evolutionary path followed by a star with initial mass of 0.5 $M_\odot$ and accreting up to a final mass of 1.2 $M_\odot$ is shown in Fig. 34. Although the physical effects are well understood, it remains to be seen if the assumption that most of the mass is accreted during PMS evolution is confirmed by observations.

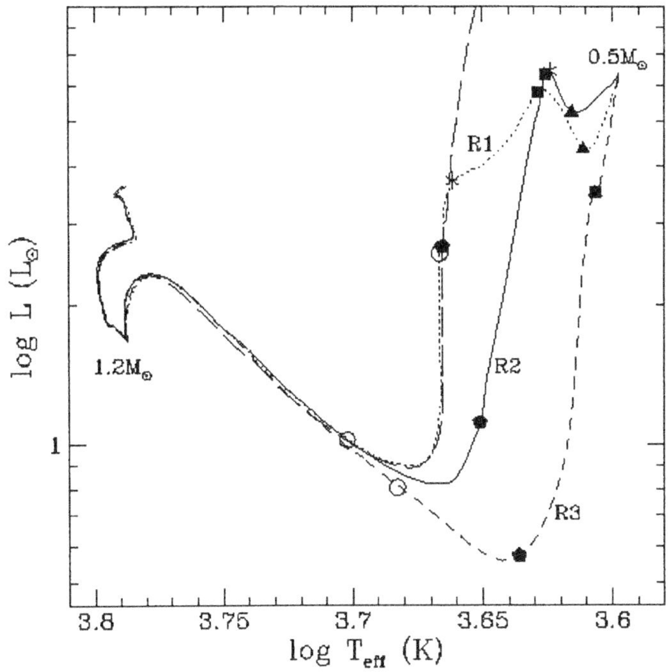

Fig. 34. Evolutionary path of a PMS accreting star. At $t = 0$, the star has a mass of 0.5 $M_\odot$, and arrives on the ZAMS with a final mass of 1.2 $M_\odot$. The various curves are for different mass accretion histories, while the non-accreting case of a 1.2 $M_\odot$ star is shown by the long-dash line. Filled triangles mark the time when the star expands as a reaction to accretion, whereas squares indicate the onset of gravitational contraction. The appearance of the radiative core is indicated by the pentagons. Finally, the end of accretion is marked by empty circles. (From Siess et al. [68])

We conclude this section by discussing an interesting observational test for the occurrence of protracted accretion in PMS stars based on the behavior of the surface abundances of deuterium and lithium (see Siess et al. [69]). Deuterium is expected to be completely destroyed in the interior of stars less massive than ~1.2 $M_\odot$ since the temperature at the base of the convection zone always remains above $\sim 10^6$ K. However, if accretion persists at a rate faster than the nuclear burning rate, then its surface abundance can build up and reach the initial interstellar value. Similarly, an increase in the abundance of surface lithium can be achieved in stars with mass $\lesssim$1.5 $M_\odot$ which, in the absence of accretion, would deplete it. As an illustration, Fig. 35 shows the domains in the H-R diagram where accretion enhances the surface abundances of both deuterium and ^{7}Li. The detection of these elements in PMS stars of the appropriate mass would provide a clear signature for long-lasting accretion episodes.

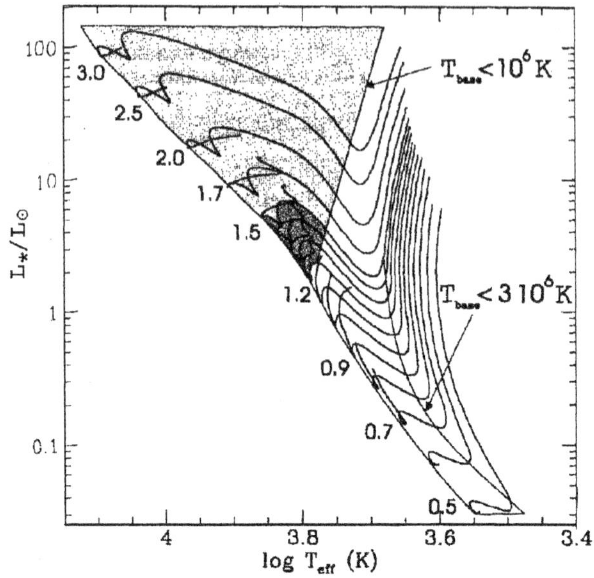

Fig. 35. Regions of the H-R diagram with enhanced surface abundance of D and ^{7}Li due to residual accretion. The shaded regions delimit the conditions at the bottom of the convection zones appropriate for D-burning ($T_{\rm base} < 10^6$ K) and Li-burning ($T_{\rm base} < 3 \times 10^6$ K). (From Siess et al. [69])

9 Rotation and PMS Evolution

9.1 Rotational Properties of Young Stars

The evolution of angular momentum in the PMS phase is a long-standing problem in the study of early stellar evolution and star formation in general. During star formation, almost all the angular momentum should be removed from the molecular cloud core. If the velocity gradient of 0.3–2 km s^{-1} pc^{-1} observed in molecular lines toward some dense ammonia cores is due to rotation, the specific angular momentum reaches values $\sim 3 \times 10^{21}$ cm^2 s^{-1} $(R/0.1\ {\rm pc})^2$ $(\Omega/4\ {\rm km\ s^{-1}\ pc^{-1}})$. On the other hand, the typical rotation periods and radii of classical T Tauri stars are P =3–10 days and $R_* \simeq 2\ R_\odot$ (Bouvier et al. [12]) which yield a value of the specific angular momentum of approximately $j \simeq 4\pi R_*^2/(5P) \sim 6 \times 10^{16}$ cm^2 s^{-1} $(R_*/2\,R_\odot)^2$ $(P/10\,{\rm days})^{-1}$ (assuming uniform rotation). Thus, the specific angular momentum should decrease by several orders of magnitude during star formation. Various mechanisms have been proposed to solve this basic problem, including binary formation, magnetic braking, magnetocentrifugally driven outflows. However, there is no consensus on which one is the dominant process.

The youngest optically visible T Tauri stars are typically rotating at only $\lesssim$1/10 of their breakup velocity (Hartmann et al. [30]; Bouvier et al. [11]).

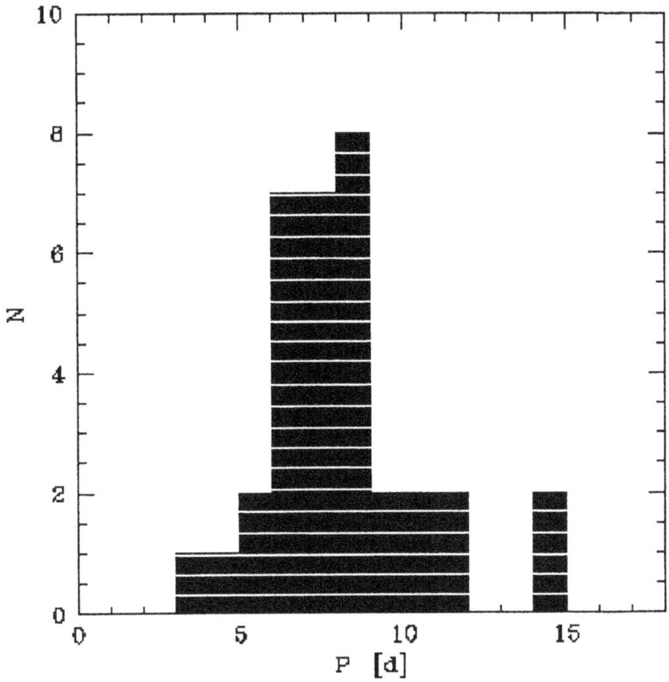

Fig. 36. The distribution of rotation periods for 34 classical T Tauri stars. (From Bouvier et al. [13])

The distribution of rotational periods of a sample of classical T Tauri stars is shown in Fig. 36. Note the well defined peak around 7–9 days (Bouvier et al. [13]). The intermediate mass Herbig Ae/Be stars also rotate at a small fraction of the breakup speed (e.g. Finkenzeller [23]). Even in the youngest deeply embedded stars the rotational rates are low (Eiroa & Casali [17]; Montmerle et al. [52]). In young clusters, such as the Orion Nebula Cluster, about 2/3 of the low mass members are slow rotators, while the rest have higher rates (Choi & Herbst [16]). Finally, young stars are expected to spin up considerably during their PMS phase, both as a result of contraction and of residual accretion from circumstellar disks. However, the rotation rates of T Tauri stars projected to the main sequence under the assumption of angular momentum conservation do not reproduce the observed rates of ZAMS stars, which are predominantly slow rotators (e.g. Queloz et al. [60]).

The older post-T Tauri stars with age $\gtrsim 10^7$ yr are in general fast rotators, with only ~30% having velocities $\lesssim 20$ km s^{-1} (Bouvier et al. [13]). The fraction of rapid rotators decreases rapidly with stellar ages: in the young open cluster α Per (50×10^6 yr) about half of the stars have projected velocities $v sini \gtrsim 50$ km s^{-1}, with a rather large spread. In the Pleiades (80×10^6 yr), about 80% of the stars are slow rotators. Finally, in the Hyades (600×10^6

yr) all stars rotate at velocities less than 10 km s^{-1}. Thus, after arrival on the main sequence, low-mass stars undergo an efficient braking via a magnetically driven wind (e.g. Weber & Davis [75]). Wind braking, although certainly present even during the initial phases of contraction, is not expected to dominate the rotational evolution since it operates on a time scale comparable or longer than the Kelvin-Helmholtz time (Bouvier et al. [10]).

Presently, the most popular mechanism for the depletion of angular momentum involves magnetic star-disk interaction (Ghosh & Lamb [25]; Königl [38]; Shu et al. [65]). According to this model, the stellar magnetic field threads the circumstellar disk and truncates it at a characteristic radius set by the balance between the accretion rate and the strength of the magnetic field. Accretion of disk material onto the star occurs along field lines, producing hot spots near the magnetic poles. At the same time, magnetic torques transfer angular momentum away from the star to the disk. The model accounts for the typical rotation period of a T Tauri star or a slow rotator in the Orion cluster (∼8 days) for reasonable values of the magnetic field strengths, few hundred Gauss, and accretion rates of the order of 10^{-11} to 10^{-9} $M_{\odot}$ yr^{-1} (e.g. Armitage & Clarke [1]).

Indirect evidence of the link between rotation rates and disk indicators is provided by the infrared and ultraviolet excesses, outflow properties, and truncated disks. For example, Edwards et al. [21] have observed that of 34 late-K and M stars in Taurus-Auriga, only those with $P > 4$ days showed significant near-infrared emission indicative of circumstellar disks. Similarly, the distribution of rotation periods among 75 stars in the Orion cluster appears to be bimodal and dividing at a period of 4–5 days, suggesting the presence of two distinct regimes of slow ($P > 4$ days) and fast ($P < 4$ days) rotators (Choi & Herbst [16]). Finally, Bouvier et al. [12] have shown that classical T Tauri stars rotate more slowly on average than the weak-line T Tauri counterparts, implying that the presence of disks affects the rotational properties of the central stars. The whole observational picture provides the basis of the "disk-locking" hypothesis.

Simplified models of the rotational evolution of low-mass stars in this framework have been developed by Bouvier et al. [10] and Siess & Livio [67]. The basic assumption is that, as long as a star is coupled to its circumstellar disk, its angular velocity is constant and fixed at the initial value. At later times, when the disk disappears, the star rotates either as a solid body (Bouvier et al. [10]) or it experiences a core-envelope decoupling at different rotation rates (Siess & Livio [67]). Obviously, the disk life time is a free parameter of the model. In all cases, the angular momentum J is extracted from the surface by a magnetized wind whose functional form is parametrized as (Kawaler [35]):

$$\frac{dJ}{dt} \sim \Omega^{1+\frac{4an}{3}} R_*^{2-n} \dot{M}_{\rm loss}^{1-\frac{2n}{3}} M_*^{-n/3} \,, \tag{65}$$

where $n = 1.5$ corresponds to a field geometry intermediate between dipolar and radial, and a describes the dependence of the magnetic field on the rotational velocity. The angular momentum loss rate is specified in two cases where $a = 0$, when the angular velocity exceeds the critical value for saturation (ω_{sat}), and $a = 1$ when the dynamo generated magnetic field is proportional to the angular velocity:

$$\left(\frac{dJ}{dt}\right)_\omega \sim -K\,\Omega^3 \left(\frac{R_*}{R_\odot}\right)^{1/2} \left(\frac{M_*}{M_\odot}\right)^{-1/2} \quad (\Omega^2 < \omega_{sat}) \tag{66}$$

and

$$\left(\frac{dJ}{dt}\right)_\omega \sim -K\,\Omega^2\,\omega_{sat} \left(\frac{R_*}{R_\odot}\right)^{1/2} \left(\frac{M_*}{M_\odot}\right)^{-1/2} \quad (\Omega \gtrsim \omega_{sat}) \tag{67}$$

For ages greater than the disk life time, the evolution of the angular velocity is then described by

$$\frac{1}{\Omega}\frac{d\Omega}{dt} = \frac{1}{J}\left(\frac{dJ}{dt}\right)_\omega - \frac{1}{I}\frac{dI}{dt} \tag{68}$$

where $J = I\Omega$ and $I = k^2 M_* R_*^2$ is the stellar moment of inertia.

The results for the rotational evolution of a 1 $M_\odot$ star is shown in Fig. 37, for an initial rotation period of 8 days and $\omega_{sat} = 14\ \Omega_\odot$ ($\Omega_\odot = 2.9 \times 10^{-6}$ s^{-1}; Bouvier et al. [10]). The behavior of $\Omega(t)$ is very sensitive to the assumed disk life time: if this is short ($\lesssim 10^6$ yr), then high angular velocities can be reached by the age of young clusters. Conversely, the lowest observed velocities can result from long lived disks ($\gtrsim 10^6$ yr). Bouvier et al. conclude that the observed distributions of projected rotational velocities can be well reproduced assuming solid body rotation and a median disk life time of $\sim 3 \times 10^6$ yr. While the models account for the moderate and fast rotators on the main sequence, the relatively high proportion of slow rotators ($\lesssim 10$ km s^{-1}) cannot be explained. It is possible that for these objects one should relax the assumption of solid body rotation and consider instead differential rotation (Krishnamurthi et al. [39]) or radiative-convective decoupling (Keppens et al. [36]). As cautioned by Bouvier et al., however, progress in the understanding of the slow rotators is hampered by the absence of precise velocity measurements in young clusters.

The disk-locking scenario has been recently challenged by the observations of a large sample of 254 stars in an area centered around the Trapezium cluster (Stassun et al. [71]). The distribution of rotation periods appears statistically indistinguishable from a uniform distribution, with no evidence of bimodality around 4–5 days. Also, there is no difference for stars with and without accretion signatures, with the former occurring over the entire range of periods observed. Finally, no correlation is found between the rotation period and the near-infrared signatures of disks. Another important finding of the study is the large dispersion of rotation rates of the low-mass population of the cluster, already present at an age of only 1–2 Myr. Such a dispersion

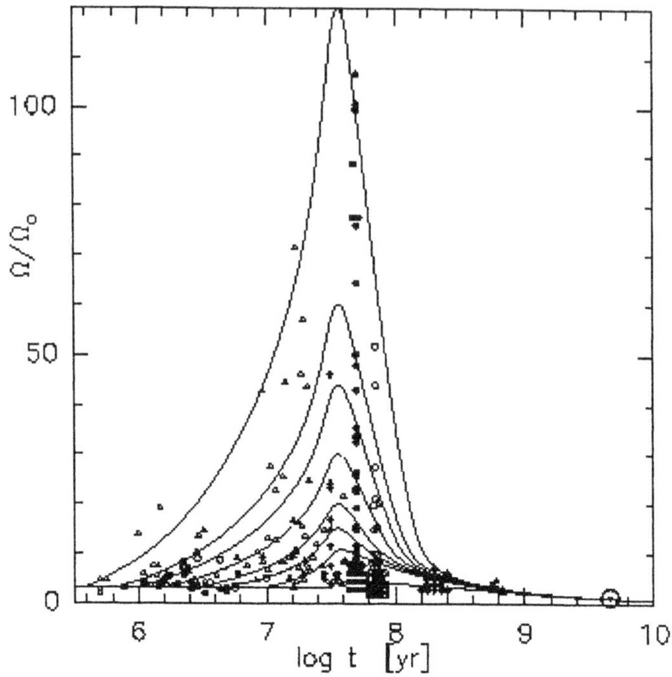

Fig. 37. The evolution of the angular velocity Ω for a 1 $M_{\odot}$ star as a function of time for various disk life times. The uppermost curve is for stars with short lived disks (4×10^5 yr), while the lowest one is for long-lived disks (30×10^6 yr). The observed angular velocities in a variety of T Tauri stars and open clusters are shown by the various symbols. (From Bouvier et al. [10])

is similar to that observed among the low-mass members of the Pleiades at ~100 Myr (see Fig. 37). Whether this dispersion is a result of the initial conditions of star formation or of rapid angular momentum evolution is still an open problem that future observations (and models) of deeply embedded objects should address.

The issue of the intrinsic difference between slow and fast rotators in Orion has been reconsidered by Herbst et al. [32] who find that the rotational properties depend in fact on stellar mass. In particular, stars more massive than ~0.25 $M_{\odot}$ have a bimodal distribution with a significant gap near 4 days, whereas less massive stars have a narrower distribution with no gaps. The observational situation is illustrated in Fig. 38 that displays the period distribution of 52 stars more massive than 0.25 $M_{\odot}$ in the Stassun et al. sample that fall within the area covered by Herbst et al. The distribution appears bimodal and the origin of the gap is related to the rapid evolution of young stars that have contracted from the birthline conserving angular momentum, thus supporting the idea of disk-regulated rotation at least for the more massive objects. However, searches in the near- and mid-infrared for circumstellar disks around stars with known rotation periods have yielded largely negative results (Stassun et al. [72]). On the other hand, when the disks are present they extend beyond the co-rotation radius, in contrast to the requirements of disk regulated angular momentum evolution. Thus, we conclude the direct linkage between stellar rotation and disks has still to be demonstrated.

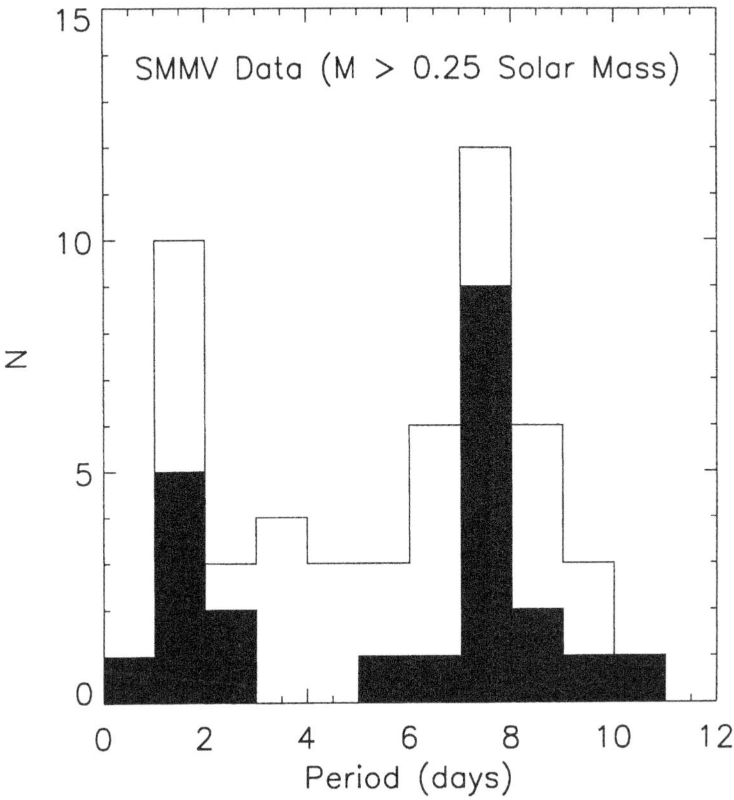

Fig. 38. The frequency distribution of stars with $M>0.25\ M_\odot$ and rotation periods determined by Stassun et al. [71]. The solid portion of the histogram is for stars whose rotation periods are confirmed in the study by Herbst et al. [32], suggesting the reality of the gap. (From Herbst et al. [32])

9.2 Structural Effects

The fact that young stars rotate faster than more evolved objects implies that one cannot neglect the effect of rotation in realistic PMS evolutionary models. This task was initiated by Kippenhan & Thomas [37] who devised a scheme to calculate the structural effects on spherically symmetric models. The method has been implemented by Endal & Sofia [22] and Pinsonneault et al. [59] to follow the PMS contraction of solar-type stars.

Recently, Sills & Pinsonneault [70] have extended the mass range to include stars less massive than 0.5 $M_\odot$ for which the observational data on the rotation rates as a function of age are abundant. The basic modifications in the standard stellar structure equations (see Sect. 4.2) consists in computing the physical variables on equipotential surfaces rather than spherical surfaces. The mass continuity and energy equations retain their non-rotating form, while the equations of hydrostatic equilibrium (eq. (33)) and radiation transport (eq. (35)) are modified to account for the change of the gravitational potential on the non-spherical equipotential surfaces. The former becomes:

$$\frac{dP}{dM_r} = -\frac{G\,M_r}{4\,\pi\,r^4}\,f_P\,, \tag{69}$$

where

$$f_P = \frac{4\,\pi\,r^4}{G\,M_r\,\mathcal{S}}\,\frac{1}{\langle g^{-1}\rangle}\,, \tag{70}$$

$$\langle g^{-1}\rangle = \frac{1}{\mathcal{S}}\int_{\psi=const} g^{-1}\,d\sigma\,. \tag{71}$$

The diffusion equation can be rewritten as:

$$\frac{dlnT}{dlnP} = \frac{3\,\kappa\,P\,L_{\rm int}}{16\,\pi\,a\,c\,G\,T^4}\,\frac{f_T}{f_P}\,, \tag{72}$$

where

$$f_T = \left(\frac{4\,\pi\,r^2}{\mathcal{S}}\right)^2\,\frac{1}{\langle g\rangle\,\langle g^{-1}\rangle}\,, \tag{73}$$

and $\langle g\rangle$ is analagous to $\langle g^{-1}\rangle$. f_T is of order unity.

The resulting evolutionary tracks for stars with mass between 1 $M_\odot$ and 0.1 $M_\odot$ are shown in Fig. 39. All models start at the birthline with an initial period of 8 days, corresponding to the median value of the CTTS. The main effect of rotation is to shift the tracks to lower effective temperatures and luminosities, mimicking a star of lower mass (Sackmann [64]). The change is more conspicuous for more massive stars and becomes neglible below ~0.5 $M_\odot$. However, since most low-mass stars have rotational velocities smaller than about 20 km s^{-1}, the net effect on the reduction of the effective temperature is always limited to few tens of degrees. Only for fast rotators ($V_{\rm rot}$ ~100 km s^{-1}) the temperature can decrease by about 100 K.

The results show only a weak dependence on the exact form of transport of internal angular momentum. Fully convective stars are assumed to rotate as solid bodies, whereas stars with radiative cores have differential rotation. All low-mass stars begin their contraction as fully convective objects, but the

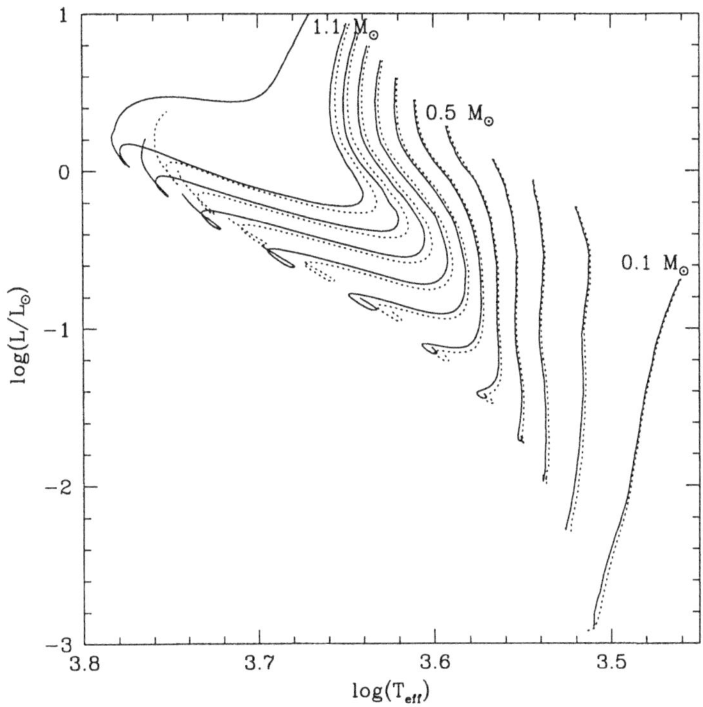

Fig. 39. The effect of rotation on evolutionary tracks. The solid lines are for non-rotating stars, and the dotted lines are stars with initial rotation periods of 8 days and no angular momentum loss. Computed masses go from 0.1 to 1.0 $M_\odot$, in steps of 0.1 $M_\odot$. (From Sills & Pinsonneault [70])

more massive objects develop radiative cores and become more centrally condensed so that the central core spins up more than the envelope. Therefore, the effects of rotation are more pronounced for differential rotators than for solid body rotators with the same surface rotation rate. However, the global modification of the stellar tracks is minor. Although still relatively simple (no angular momentum loss, no internal instabilities), these models represent an interesting assessment of the impact of rotational effects on PMS contraction. It should be pointed out, however, that the models do not address the basic angular momentum problem related to the low rotational velocities observed in the majority of the young stars.

10 Pulsational Instabilities in PMS Stars

Young stars are characterized by a large degree of activity. Winds, jets and outflows are manifestations of the interaction of the stars with the circumstellar environment in which they are embedded. Similarly, both T Tauri and Herbig Ae/Be stars show photometric and spectroscopic variability on time scales of minutes to years, indicating that photospheric activity begins in the earliest phases of stellar evolution, prior to the arrival on the Main Sequence (MS). In many cases, strong and rapid line variability has been detected with periods close to the star rotation period. This variability has been interpreted as due to the rotational modulation of spots on the stellar surface or of fast and slow streams in the stellar wind (Catala et al.[18]). On the other hand, the fact that young stars during their evolution to the MS move across the instability region of post-MS stars raises the possibility that at least part of the activity could also be due to stellar pulsations (Baade & Stahl [2]; Kurtz & Marang [40]). However, it is not at all clear whether a PMS star can indeed pulsate.

As emphasized by Gautschy & Saio [24], stars over essentially the whole mass spectrum can become pulsationally unstable during various stages of their evolution. In particular, δ Scuti stars are intermediate-mass variables of spectral type A to F with pulsation periods less than $0\overset{d}{.}3$ and light amplitudes ranging from thousandths of magnitude to some tenths (Breger 1979 [14]). In spite of the traditional association of δ Scuti pulsation with stars in their core hydrogen-burning phase or evolving towards the base of the giant branch and burning hydrogen in a shell, some δ Scuti have been identified to be PMS objects. In particular, the Herbig Ae star HR 5999 has shown photometric light variations characteristic of δ Scuti stars with peak-to-peak pulsation amplitude of about $0\overset{m}{.}013$ in Johnson V-band and a period of 4.99 hr (Kurtz & Marang [40]). These small amplitude changes are superimposed on a non-periodic photometric variation of $0\overset{m}{.}35$, probably due to variations of dust obscuration in the circumstellar envelope and disk.

Of particular interest is also the Ae star HD 104237 that shows both short- and long-term velocity changes of spectral lines (Donati et al. [20]). These variations indicate that the star is undergoing radial pulsations with a period of approximately 40 minutes and an amplitude of about 1 km s^{-1}. Interestingly, HD 104237 is the first intermediate mass PMS star with a measured magnetic field .

The detection and characterization of the pulsation properties of young stars provides an opportunity to improve the knowledge of their internal structure and to obtain unique constraints on the theoretical predictions of the models. Ultimately, the analysis of the pulsation characteristics can yield an indirect estimate of the stellar mass. This represents a powerful method for stars that are not part of the restricted group of spectroscopic binary systems (see Sect. 10.2).

10.1 The Input Models

The calculation of the pulsational instability requires the specification for any given mass of the stellar luminosity and effective temperature. These are provided by PMS models as a function of time. The physical mechanisms which govern the pulsation phenomenon are the κ and γ mechanisms in the hydrogen and helium ionization regions (see Cox [19]). Thus, one must rely on the radiative opacities, such as OPAL, for temperatures higher than 10^4 K, and molecular opacities for lower temperatures (e.g., Alexander & Ferguson [1]).

As an illustration of the differences in the internal structure of stars at various stages of evolution, Fig. 40 shows the variation with radius of the interior density (normalized to the central value) for a 3 $M_\odot$ star. Three cases are considered: a protostar, a young PMS star, and an evolved object. On the birthline, the density profile is quite flat and the star is thermally unrelaxed. At the end of the relaxation phase, in the middle of the PMS track, the star has already started its contraction to the ZAMS and is more centrally condensed. The middle curve of Fig. 40 is for a model with luminosity L_*=57 $L_\odot$ and effective temperature $T_{\rm eff}$=6900 K at an age of 7×10^5 yr. As we shall see below, such a model falls within the boundaries of the instability strip. Finally, the lower curve is for a star that has completed the main-sequence phase and is evolving toward the red giant branch. The effective temperature is approximately the same as for the PMS star, but the luminosity is now L_*=135 $L_\odot$ due to the larger radius (8.3 $R_\odot$ instead of 5.3 $R_\odot$). These conditions correspond to those of a typical, evolved δ Scuti star. The star is much more centrally condensed because of the complete exhaustion of hydrogen in the center. However, beyond the inert core, the density profile is quite similar to that of the PMS star. Thus, a mature δ Scuti star differs from a young one mostly in the inner regions and for its higher luminosity. These structural changes clearly affect the estimate of the pulsation periods, as we shall see in the following sections.

10.2 The Instability Strip

The theoretical framework for the investigation of the linear and nonlinear pulsation characteristics is well known (e.g. Bono et al. [9]). For each selected pulsation mode, the fundamental observables that can be derived from the linear nonadiabatic computations are the periods and growth rates. A detailed analysis of the sign of growth rates provides an estimate of the location of the blue boundaries of the instability strips for each mode, representing the maximum effective temperatures allowing pulsation at a given luminosity level. No information is derived about the location of red boundaries because the linear models are completely radiative while the quenching of pulsation at lower effective temperatures is due to the interaction between convective and dynamical motions (e.g. Baker & Kippenhan [3]).

Given the static envelope structures produced by the linear computations, one can then perform the nonlinear stability analysis. The equations governing both the dynamical and convective structures must be integrated in time until the initial perturbations and the nonlinear fluctuations, which result

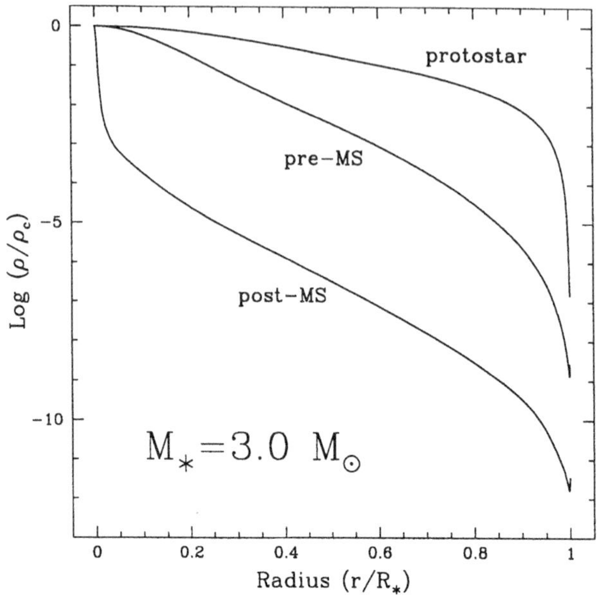

Fig. 40. The internal density profile (normalized to the central value) of a 3 $M_\odot$ star in different evolutionary stages. The *upper curve* is a protostar on the birthline computed with $\dot{M}_{acc} = 10^{-5}\ M_\odot\ yr^{-1}$. The *middle curve* and *lower* curves are for a pre-MS and for a post-MS star with similar effective temperatures but different luminosities. (From Marconi & Palla [43])

from the superposition of higher order modes, settle down. Because of the very low growth rates (between 10^{-5} and 10^{-4}), an extremely large period number (exceeding 10,000) must be calculated for each model before reaching the stable limit cycle of pulsation (maximum amplitude).

The modal stability analysis reveals the presence of stable nonlinear limit cycles in the fundamental, first and second overtone as the effective temperature of the star increases. Figure 41 shows the location of the theoretical instability strip boundaries in the H-R diagram, together with the last static and the first pulsating models for various masses (Marconi & Palla [43]). The stellar parameters that define the strip are listed in Table 3. The first and second overtone pulsators are located at effective temperatures higher than the fundamental ones. All the models on the red edge are pulsating in the fundamental mode, while those on the blue edge in either of the overtone modes. The *red* edge varies between ~6500 K and ~7100 K, whereas the *blue* edge between ~7100 *K* and ~7500 K for different masses. The width of the instability strip is nearly constant and equal to about 650 K. The time spent by each star within the strip is also a constant fraction of the total PMS contraction time, typically between 5 and 10%, although in absolute terms it decreases from ~10^6 yr for M_*=1.5 $M_\odot$ to 8×10^4 yr for M_*=4 $M_\odot$.

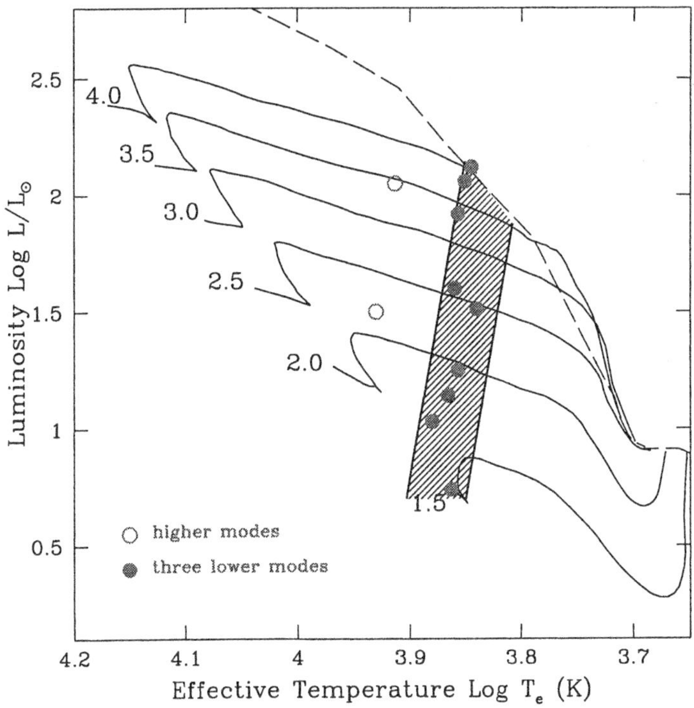

Fig. 41. The shaded region shows the location of the instability strip of PMS stars in the H-R diagram. The various symbols indicate the distribution within or near the boundaries of the instability strip of Herbig Ae stars with evidence for pulsational variability. (From Marconi et al. [46])

As shown in Fig. 41, the PMS instability strip terminates abruptly at the blue edge of the $M_*=4\ M_\odot$ evolutionary track. This results from the assumption that PMS stars begin their contraction phase at the birthline computed with an accretion rate of $\dot{M}_{acc} = 10^{-5}\ M_\odot\ yr^{-1}$. Obviously, the theoretical instability strip extends to luminosities higher than the upper limit shown here for PMS stars, but it would then be populated only by more evolved objects such as the classical δ Scuti stars.

Despite the relatively short time spent by PMS stars inside the strip, Fig. 41 shows that several Herbig Ae stars with the right combination of luminosity and temperature have been identified as candidate pulsators. In addition to HR 5999 and HD 104237 mentioned before, δ Scuti type variations have been discovered in HD 35939 and V351 Ori (Marconi et al. [44], [45]), in BL 50 and HP 57 members of the open cluster NGC 6823 (Pigulski et al. [57]), and HD 142666 (Kurtz & Müller [41]). The comparison between the observed pulsation properties and the predictions of non-linear hydrodynamical models

Table 3. Parameters of the instability strip

	Red Edge				Blue Edge			
Mass ($M_\odot$)	L_* ($L_\odot$)	T_{eff} (K)	Mode	Period (d)	L_* ($L_\odot$)	T_{eff} (K)	Mode	Period (d)
1.5	7	7090	f	0.0611				
2.0	18	6870	f	0.1195	21	7500	II	0.0690
2.5	32	6640	f	0.2155	37	7350	II	0.0925
3.0	52	6500	f	0.2601	62	7260	II	0.1299
3.5	77	6550	f	0.3195	95	7200	II	0.1721
4.0					133	7100	I	0.2736

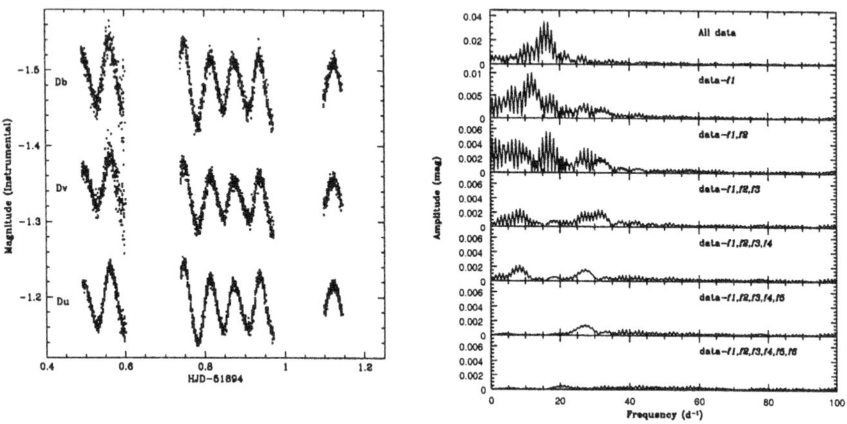

Fig. 42. Left panel: *UBV* photometry of V351 Ori. The *B* and *U* data have been shifted by ±0.07 mag for clarity. Also, the HJD for the second and third night are displaced by −0.65 day and −1.3 day, respectively. **Right panel**: Frequency analysis in the *U* filter. The first diagram shows the Fourier Transform of all the data, while the other panels show the subtraction of the derived frequencies. (From Marconi et al. [45])

provides important constraints on the mass and evolutionary state of the Herbig stars.

A case in point is V351 Ori. Figure 42 shows the light curves in the *UBV* Johnson system obtained during three nights in December 2000. Light variations are presnet in each filter with a maximum amplitude of ∼0.12 mag in *U* and *B*. The frequency analysis reveals the presence of multi-periodic oscillations, as displayed in the right panel of the figure. Up to four frequencies have been identified with periods between 0.03 and 0.08 days, indicating that V351 Ori is pulsating in a mixture of modes. The pulsation analysis

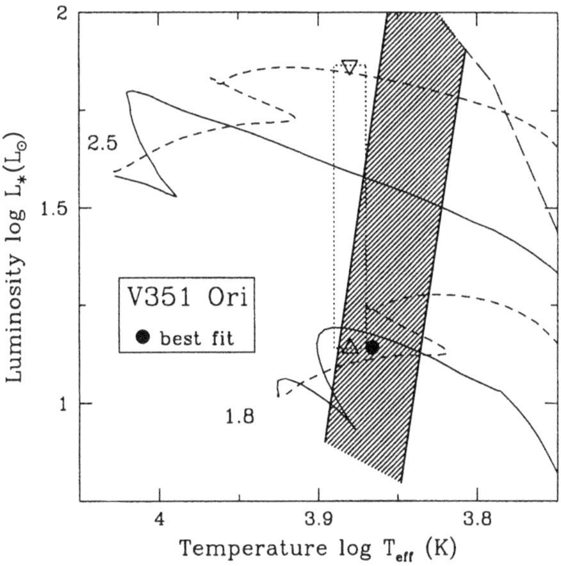

Fig. 43. The position of V351 Ori in the H-R diagram with the uncertainty on spectral type and distance indicated by the box. The instability strip is shown by the shaded region. The PMS and post-MS evolutionary tracks are shown for 1.8 $M_\odot$ and 2.5 $M_\odot$ (heavy and thin lines, respectively). The filled circle marks the position of the best fit model from the pulsation analysis. (From Marconi et al. [45])

successfully matches the detected frequencies with a single model with mass $M_* = 1.8M_\odot$, luminosity $L* = 13.9L_\odot$, and effective temperature $T_{\rm eff} = 7350$ K. The position of the best fit model in the H-R diagram is shown in Fig. 43. The dotted box accounts for the uncertainty on the spectral type and distance (d =260 to 450 pc; van den Ancker et al. [50]). The lower value is the Hipparcos value, whereas the upper limit assumes that the star is located in the Orion star forming region. The best fit model suggests a mass of 1.8 $M_\odot$ at the Hipparcos distance, indicating that the star is not associated with Orion.

The position of V351 Ori in the H-R diagram does not rule out the possibility that the star is an evolved object rather than a young PMS star. According to Fig. 43), the post-MS track of a 1.8 $M_\odot$ star intersects the corresponding PMS track within the instability strip, close to the location of V351 Ori. Since the pulsational properties depend only on the surface properties (which are identical for pre- and post-MS stars), there is no clear signature that would allow to distinguish between a young (~6.5 Myr) and an evolved (~1 Gyr) star. However, this is a classical situation where asteroseismological studies can offer an unambiguous answer on the evolutionary stage of the star (and similar cases). In fact, the internal structure of an evolved star differs markedly from that of a young one and the presence of mean molecular weight gradients in the former can induce the excitation of *non radial* modes (Suran et al. [74]). The expected amplitudes are large enough for detection with space telescopes, such as the COROT satellite. Thus, all the PMS stars that fall inside the instability strip of Fig. 41 are excellent candidates for asteroseismological observations.

References

1. Armitage, P.J., Clarke, C.J. 1996, MNRAS, 280, 458
2. Baade, D., Stahl, O. 1989, A&A, 209, 268
3. Baker, N.H., Kippenhan, R. 1965, ApJ, 142, 868
4. Barrado y Navascués, D., Stauffer, J.R. 1996, ApJ, 310, 879
5. Basri, G., Martín, E.L. 1999, AJ, 118, 2460
6. Basri, G., Marcy, G.W., Graham, J.R. 1996, ApJ, 458, 600
7. Bildsten, L. Brown, E.F., Matzner, C.D., Ushomirsky, G. 1997, ApJ, 482, 442
8. Bodenheimer, P. 1965, ApJ, 142, 451
9. Bono, G., Caputo, F., Cassisi, S., Castellani, V., Marconi, M., Stellingwerf, R.F. 1997 ApJ, 477, 346
10. Bouvier, J., Forestini, M., Allain, S. 1997b, A&A, 326, 1023
11. Bouvier, J., Bertout, C., Benz, W., Mayor, M. 1986, A&A, 165, 110
12. Bouvier, J., Cabrit, S., Fernandez, M., Martín, E.L., Matthews, J. 1993, A&A, 101, 495
13. Bouvier, J., Wichmann, R., Grankin, K., Allain, S., Covino, E., Fernandez, M., Martín, E.L., Terranegra, L., Catalano, S., Marilli, E. 1997a, A&A, 318, 495
14. Breger, M. 1979, PASP, 91, 5
15. Burrows, A., Marley, M., Hubbard, W., Lunine, J., Guillot, T., Saumon, D., Freedman, R., Sudarsky, D., Sharp, C. 1997, ApJ, 491, 856
16. Choi, P., Herbst, W. 1996, AJ, 111, 283
17. Casali, M.M., Eiroa, C. 1996, A&A, 306, 427
18. Catala, C., Simon, T., Praderie, F., Talavera, A., Thè, P.S., Tjin A Djie, H.R.E. 1989, A&A, 221, 273
19. Cox, J.P. 1980, Theory of Stellar Pulsations, Princeton University Press
20. Donati, J.F., Semel, M., Carter, B.D., Rees, D.E., Cameron, A.C. 1997, MNRAS, 291, 658
21. Edwards, S., Strom, S.E., Hartigan, P., Strom, K.M., Hillenbrand, L.A., Herbst, W., Attridge, J., Merrill, K.M., Probst, R., Gatley, I. 1993, AJ, 106, 372
22. Endal, A.S., Sofia, S. 1976, ApJ, 210, 184
23. Finkenzeller, U. 1985, A&A, 151, 340
24. Gautschy, A., Saio, H. 1996, ARAA, 34, 551
25. Ghosh, P., Lamb, F.K. 1979, ApJ, 232, 259
26. Gullbring, E., Hartmann, L., Briceño, C. & Calvet, N. 1998, ApJ, 492, 323
27. Hartigan, P., Edwards, S., Ghandour, L. 1995, ApJ, 452, 736
28. Hartmann, L., Kenyon, S.J. 1990, ApJ, 349, 190
29. Hartmann, L., Calvet, N., Gullbring, N., D'Alessio, P. 1998, ApJ, 495, 385
30. Hartmann, L.W., Hewett, R., Stahler, S.W., Mathieu, R.D. 1986, ApJ, 309, 275

31. Herbig, G.H. 1978, in Problems of Physics and Evolution of the Universe, ed. L.V. Mirzoyan (Erevan: Acad.l Sci. Armenian SSR), p. 171
32. Herbst, W., Rhode, K.L., Hillenbrand, L.A., Curran, G. 2000, AJ, 119, 261
33. Hillenbrand, L.A., Strom, S.E., Vrba, F.J., Keene, J. 1992, ApJ, 397, 613
34. Hirth, G.A., Mundt, R., Solf, J. 1997, A&AS, 126, 437
35. Kawaler, S.D. 1988, ApJ, 33, 236
36. Keppens, R., MacGregor, K.B., Charbonneau, P. 1995, A&A, 294, 469
37. Kippenhan, R., Thomas, H.-C. 1970, in Stellar Rotation, ed. A. Slattebak (Dordrecht: Reidel), p. 20
38. Königl, A. 1991, ApJ, 370, L39
39. Krishnamurthi, A., Pinsonneault, M.H., Barnes, S., Sofia, S. 1997, ApJ, 480, 303
40. Kurtz, D.W., Marang, F. 1995, MNRAS, 276, 191
41. Kurtz, D.W., Müller, M. 2001, MNRAS, 325, 1341
42. Magazzù, A., Rebolo, R., Martín, E.L. 1992, ApJ, 392, 159
43. Marconi, M., Palla, F. 1998, ApJ, 507, L141
44. Marconi, M., Ripepi, V., Alcalá, J.M., Covino, E., Palla, F., Terranegra, L. 2000, A&A, 355, L35
45. Marconi, M., Ripepi, V., Bernabei, S., Palla, F., Alcalá, J.M., Covino, E., Terranegra, L. 2001, A&A, 372, L21
46. Marconi, M., Ripepi, V., Palla, F., 2001, in preparation
47. Martín, E.L. 1997, A&A, 321, 492
48. Martín, E.L., Claret, A. 1996, A&A, 306, 408
49. Martín, E.L., Rebolo, R., Magazzù, A., Pavlenko, Ya.V. 1994, A&A, 282, 503
50. Martín, E.L., Basri, G., Zapatero Osorio, M.R., Rebolo, R., López, R.J., Garcia, R.J. 1998, ApJ, 507, L41
51. Mendes, L.T.S., D'Antona, F., Mazzitelli, I. 1999, A&A, 341, 174
52. Montmerle, T., Grosso, N., Tsuboi, Y., Koyama, K. 2000, ApJ, 532, 1097
53. Natta, A., Grinin, V.P., Mannings, V. & Ungerechts, H. 1997, ApJ, 491, 885
54. Neuhäuser, R. 1997, Science, 276, 1373
55. Palla, F. 1994, in Infrared Astronomy, eds. A. Mampaso & F. Sanchez (Cambridge: Cambridge Univ. Press), p. 1
56. Parigi, G. 1992, Thesis, Univ. of Florence, unpublished
57. Pigulski, A., Kolaczkowski, Z., Kopacki, G. 2000, in IAU Colloq. 176 The Impact of Large Scale Surveys on PUlsating Star Research, eds. L. Szabados & D. Kurtz (San Francisco: ASP), p. 499
58. Pinsonneault, M.H., Kawaler, S.D., Demarque, P. 1990, ApJS, 74, 501
59. Pinsonneault, M.H., Kawaler, S.D., Sofia, S., Demarque, P. 1989, ApJ, 338, 424
60. Queloz, D., Allain, S., Mermilliod, J.-C., Bouvier, J., Mayor, M. 1998, A&A, 335, 183
61. Rebolo, R., Martín, E.L., Magazzù, A. 1992, ApJ, 389, L83
62. Rebolo, R., Magazzù, A., Martín, E.L. 1995, in Stellar and Interstellar Lithium, eds. F. Spite and R. Pallavicini, Mem.S.A.It., 66, 375
63. Rebolo, R., Martín, E.L., Basri, G., Marcy, G.W., Zapatero Osorio, M.R. 1996, ApJ, 469, L53
64. Sackmann, I.J. 1970, A&A, 8, 76
65. Shu, F.H., Najita, J., Ostriker, E.C., Wilkin, F., Ruden, S., Lizano, S. 1994, ApJ, 429, 781
66. Siess, L., Forestini, M. 1996, A&A, 308, 472

67. Siess, L., Livio, M. 1997, ApJ, 490, 785
68. Siess, L., Forestini, M., Bertout, C. 1997, A&A, 308, 472
69. Siess, L., Forestini, M., Bertout, C. 1999, A&A, 342, 480
70. Sills, L., Pinsonneault, M.H., Terndrup, D.M. 2000, ApJ, 534, 335
71. Stassun, K.G., Mathieu, R.D., Mazeh, T., Vrba, F.J. 1999, AJ, 117, 2941
72. Stassun, K.G., Mathieu, R.D., Vrba, F.J., Mazeh, T., Henden, A. 2001, AJ, 121, 1003
73. Stauffer, J.J.R., Schultz, G., Kirkpatrick, J.D. 1998, ApJ, 499, L199
74. Suran, M., Goupil, M., Baglin, A., Lebreton, Y. Catala, C. 2001, A&A, 372, 233
75. Weber, E.J., Davis, L. 1967, ApJ, 148, 217

IV. PMS Evolution of Clusters and Associations

The conventional theory of protostellar infall successfully accounts for the formation and early stellar evolution of low- and intermediate-mass stars up to $\sim$10–15 $M_\odot$. More massive protostars burn hydrogen while in the accretion phase and therefore join the main sequence early on, implying that stars of even higher mass have no contracting PMS phase. As discussed in previous sections, the location of the stellar birthline and the distribution of the observed T Tauri and Herbig Ae/Be stars are in agreement with these theoretical predictions. Also, observations of young clusters such as NGC 6611 and the dense regions of the Trapezium cluster support these findings. However, the accretion scenario of isolated protostars fails to explain the existence of massive stars. Radiation pressure by photons produced at the stellar and disk surfaces on the infalling matter begins to being significant at about the critical mass of 10–15 $M_\odot$. This limit can be increased by considering variations in the dust properties (abundance and size), or in the mass accretion rate. But in either case the required conditions are so extreme (dust depletion by at least one order of magnitude and accretion rates in excess of $\sim 10^{-3}$ $M_\odot$ yr $^{-1}$) that cannot reasonably apply in all circumstances. Disk accretion can certainly help to shift the critical mass to higher values (e.g., Nakano [69]).

Our current understanding of the physics of star formation is restricted to conditions that best apply to the buildup of individual protostars within dense cores and the effect of the environment is often neglected. This is an oversimplification even in the case of Taurus-Auriga, which is considered the prototypical region for isolated star formation, where the the majority of the young stellar population is in binary or multiple systems (e.g., Ghez et al. [31]; Leinert et al. [58]). A major progress in observational studies of star formation comes from the evidence that most stars form primarily in groups. The degree of richness varies from aggregates of few tens of stars, as in the case of the T Tauri stars in the Taurus-Auriga complex (e.g., Gomez et al. [33]), to the most populous clusters in OB associations containing thousands of stars (e.g., Hillenbrand [23]). The same tendency is also found in the younger population of highly embedded stars, still invisible at optical wavelengths. Surveys carried out in the near-infrared in a variety of star forming regions, including ρ Ophiuchi, Chamaeleon and L1640 in the Orion molecular cloud , have conclusively demonstrated the occurrence of clustering of young stellar objects. Since most stars are believed to form in giant molecular clouds such as Orion, this argues for clustered star formation as the primary mode of star formation. In addition, massive stars are almost exclusively connected with clusters. The evidence that the stellar mass distribution around massive stars is consistent with the field star initial mass function (IMF) suggests that the majority of stars of all masses should also be born in clusters.

These observational facts offer important clues for the understanding of the process of massive star formation. However, a number of outstanding issues are still unanswered: Do massive stars form by accretion or by stellar mergers? Are disks present around intermediate- and high-mass stars? Do massive stars form in clusters? When does the transition from isolated to clustered formation occur? Is it possible to discover a genuine massive protostar? It is the purpose of this chapter to confront with some of these problems and describe possible avenues to provide some initial answers.

11 High Mass Star Formation

The validation of the standard accretion scenario requires the identification of a high mass protostar with signatures of gravitational infall and circumstellar disk. The observational approach to the search of the youngest high mass objects was first formulated by Habing & Israel [35]: the likely candidates must have high luminosity ($L \gtrsim 10^4$ $L_\odot$), be embedded in dense circumstellar environments, and not associated with HII regions. Thus, the attention has gradually shifted from the study of HII regions to that of ultracompact (UC) HII regions, to that of molecular clumps in which UC HII are embedded. In this case, the observed luminosity would be due to the release of gravitational infall energy and not to nuclear burning. Despite many attempts, however, no clear identification of a genuine protostar has been obtained yet (e.g., [51]).

There are two major observational difficulties: the selection of good targets and the choice of a suitable tracer. Because stars form in high density regions with huge visual extinctions ($A_V \gtrsim 100$ mag), the protostellar candidates should be visible in high density molecular tracers (line or continuum) and selected on the basis of their established association with regions of high mass star formation. A powerful two-color selection criterion has been suggested by Wood & Churchwell[105] to identify massive YSOs still embedded inside molecular clouds. These sources are characterized by the largest flux densities at 100 μm, implying that the emitting dust is quite cool (T_d $\sim$30 K); their spectra peak at $\sim$ 100μm and the shape does not show an appreciable variation from source to source. The population of embedded massive stars, and their associated ultracompact (UC) HII regions, have distinctive far-infrared colors which have been used to estimate their total number and distribution in the Galaxy, the time scale of the embedded phase (typically $\sim$10-20% of the main-sequence lifetime), and the current rate of massive star formation ($\sim 3 \times 10^{-3}$ O stars yr^{-1}).

The study at high angular resolution of individual UC HII regions reveals complex morphologies since massive stars never appear alone. Different strategies have been followed in order to identify massive protostars in these fields. Some groups have observed the molecular environment in the close vicinity of UC HII regions ([72][13] with the aim of finding a protostar in the same cluster of the early type star. Alternatively, the selection is based on sources with H_2O maser emission but *without* radio continuum emission. The assumption here is that water masers are signposts of high mass star formation, while the lack of radio continuum should identify the youngest stellar objects since the UV radiation field has not had time to create an ionized region ([14]; [17]). One of the most important results that have emerged from these studies is the detection of small (~0.1 pc), dense (~10^7 cm $^{-3}$), hot ($\gtrsim$100 K) molecular clumps, the so-called "Hot Cores" (e.g. [28]). A beautiful example of the complex environment of a hot core is shown in Fig. 44 for the case of G29.96−0.02. The peak of the high density gas as traced by the $C^{34}S$(5–4) and CH_3CN(6–5) line emission is not spatially coincident with the location of the UC HII region, identified by the continuum emission at 1.3 cm. Most of the activity in the field is associated with the embedded sources in the hot core, which excites both OH and H_2O maser emission, and *not* with the massive star ionizing the circumstellar gas.

11.1 The Search for Massive Protostars

Interestingly, the first claim of the detection of infalling motions in a molecular core was found in the case of a massive star by Ho & Haschick [43] who observed an inverse P-Cygni profile (characterized by redshifted absorption and blueshifted emission) in HCO^+ towards the UC HII region W49. However, recent observations at high-angular resolution (~ 5″) have shown that the velocity field is quite complex and most likely not due to gravitational infall. In general, the search is made difficult both for the fact that the UC HII regions are usually distant (typically more than 5 kpc), and for the presence of high-velocity molecular outflows. A statistical study of 122 high-mass star forming regions made by Shepherd & Churchwell [84] has shown that indeed outflows are a common property of newly formed massive stars, similar to the case for low-mass stars.

In addition, there are indications that the embedded objects are already main sequence stars, and not true protostars. To find the initial conditions for star formation, one should therefore be looking for the *precursors* to HII regions. The search has resulted in the identification of a small sample of bright IRAS sources (L~ $10^4 - 10^5$ $L_\odot$) with the appropriate properties: one of them, IRAS 23385+6053, with $L_{bol} = 1.6 \times 10^4$ $L_\odot$ represents an excellent candidate for a massive protostar ([62]). The spectral energy distribution is shown in Fig. 45. The source is associated with a compact outflow in HCO^+ and SiO, with dimensions less than 0.05 pc, a dynamical time scale $\lesssim 7 \times 10^3$ yr, and a mass loss rate $\gtrsim 10^{-3}$ $M_\odot$ yr^{-1}. Furthermore, the ratio $L_{submm}/L_{bol} \sim 3 \times 10^{-3}$, and $M_{env}/M_* \gg 1$. All these parameters are characteristic of a class 0 object ([3]).

The absence of radio continuum from these bright sources raises an interesting problem: can accretion stop the expansion of the HII region? And if so, what is the required mass accretion rate? The minimum accretion rate required to choke-off the expanding HII region can be estimated in a simple way (e.g., Yorke [106]). Assume balance between ionizations and recombinations:

$$N_i = 4\pi\beta_2 \int_{R_\circ}^{R_S} n_e^2 \, R^2 \, dR \,. \tag{74}$$

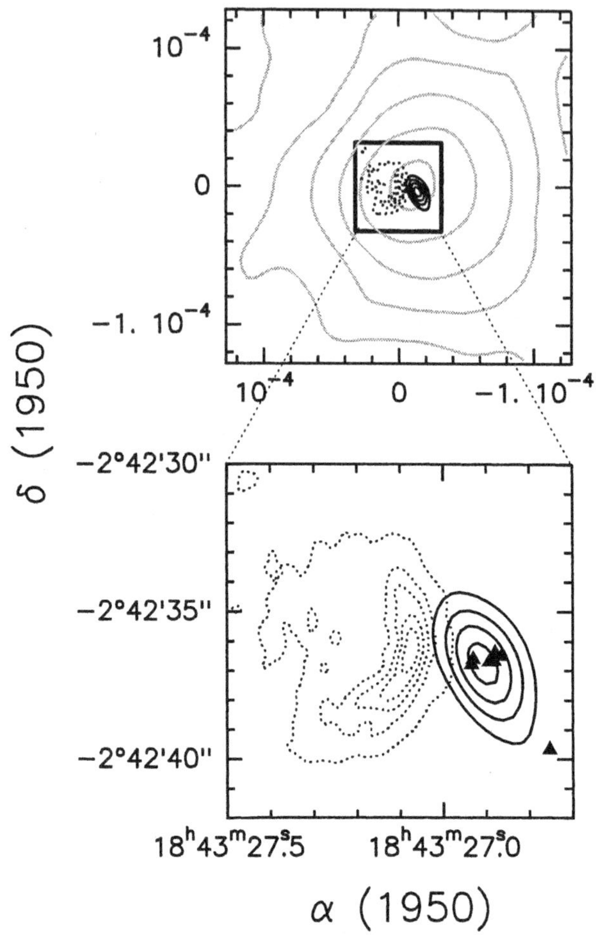

Fig. 44. The environment of the hot core associated with G29.96−0.02 Top: Comparison of the $C^{34}S$(5–4) (grey contours), CH_3CN(6–5) (black contours), and 1.3 cm continuum (dotted contours) emission toward the UC HII region G29.96−0.02. Bottom: enlargement of the region where the HC and the UC HII region are located; the triangles indicate the position of the H_2O masers. (From Kurtz[51]).

Consider a density profile of the type $n = n_\circ (R/R_\circ)^\gamma$ with $\gamma = -3/2$. Then,

$$N_i = 4\pi \, \beta \, n_\circ^2 \, R_\circ^3 \, ln \, (R_s/R_\circ) \, . \tag{75}$$

The accretion rate and free-fall velocity are given by:

$$\dot{M}_{\rm acc} = 4\pi m_H n_\circ v_{ff} R_\circ^2 \;\; ; \;\; v_{ff} = \sqrt{\frac{2GM_*}{R_\circ}} \, . \tag{76}$$

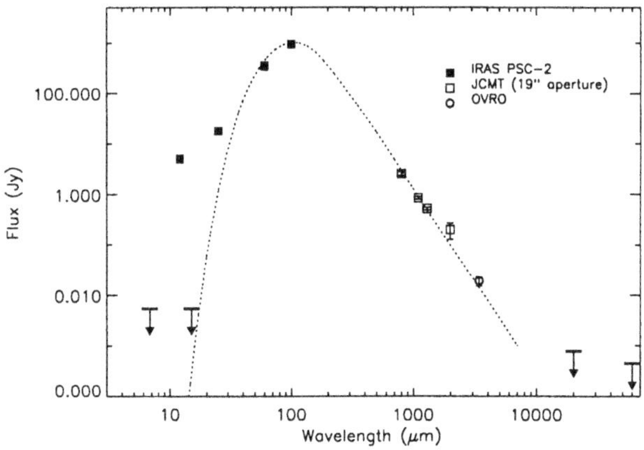

Fig. 45. Spectral energy distribution of the IRAS 23385+6053 core. Upper limits at 7 and 15 μm are from ISOCAM, those at 2 and 6 cm are from the VLA. The dotted line is a model fit. (From Molinari et al. [62])

From which we derive:

$$n_\circ^2\, R_\circ^3 \;=\; \frac{1}{2GM_*}\,\frac{1}{16\pi^2 m_H^2}\,\dot{M}_{\rm acc}^2\,. \tag{77}$$

Finally, the ratio between the Strömgren radius and the inner radius is given by:

$$\frac{R_S}{R_\circ} \;=\; exp\left[\frac{4\,\pi\, m_H^2\, N_i\, G\, M_*}{\beta_2\, \dot{M}_{\rm acc}^2}\right]\,. \tag{78}$$

Then, the condition on the critical $\dot{M}$ to avoid expansion is:

$$\dot{M}_{\rm cr} \;=\; \left(4\,\pi\, N_i\, G\, M_*\, m_H^2\, \beta_2^{-1}\right)^{1/2} \quad {\rm M_\odot yr^{-1}} \tag{79}$$

or

$$\dot{M}_{\rm cr} \sim 4\times 10^{-5}\, M_\odot\, yr^{-1} \left(\frac{N_i}{10^{49} s^{-1}}\right)^{1/2} \left(\frac{M_*}{30\, M_\odot}\right)^{1/2}\,. \tag{80}$$

Values of the critical rate are given in Table 4 for three cases. Also listed is the accretion luminosity ($L_{\rm acc} = GM_*\dot{M}_{\rm acc}/R_*$), computed for two values of the mass accretion rate, $\dot{M}_{\rm acc}$ =10^{-5} and 10^{-4} M$_\odot$ yr^{-1}, respectively. One can immediately see that, contrary to expectation, a modest mass accretion can choke-off the HII region: for an O8 star, $\dot{M}_{\rm cr} \sim 2\times 10^{-5}$ M$_\odot$ yr^{-1}, while for a B0 star $\dot{M}_{\rm cr}$ is about five times smaller. Values of few$\times 10^{-6}$ M$_\odot$ yr^{-1} are considered lower limits to the actual accretion rates in low-mass stars. Since high-mass stars should have even higher accretion rates, it would be easy to revert the expansion of the ionized region, making the sources undetectable in the radio. From Table 4 we can also note that stars more massive than $\sim$

15 $M_\odot$ have always $L_* \gg L_{acc}$, for reasonable values of $\dot{M}_{acc}$. Thus, massive YSOs may still be in the accretion phase even though the underlying source is already a main-sequence star.

Table 4. Critical mass accretion rates for massive stars

Sp. Type	M_* ($M_\odot$)	L_* ($L_\odot$)	log N_i (s^{-1})	$\dot{M}_{cr}$ ($M_\odot$ yr^{-1})	$L_{acc}(-5)$ ($L_\odot$)	$L_{acc}(-4)$ ($L_\odot$)
O5	60	8×10^5	49.7	1×10^{-4}	2×10^3	1.5×10^4
O8	23	1×10^5	48.6	1×10^{-5}	8×10^2	8×10^3
B0	18	5×10^4	47.6	4×10^{-6}	7×10^2	7×10^3
B2	10	5×10^3	43.6	6×10^{-8}	5×10^2	5×10^3

The identification of objects similar to IRAS 23385+6053 raises another fundamental question: is the observed luminosity entirely due to accretion, or does it come from nuclear energy? In the former case, we are dealing with a true protostar, whereas in the latter event the central source is nothing but a highly embedded main sequence star. In such case, the absence of radio continuum emission requires the HII region, if present, to be optically thick and very compact (~100 AU). From the table above, we see that a residual accretion rate of $\gtrsim 3 \times 10^{-6}$ $M_\odot$ yr^{-1} should be able to prevent the formation and expansion of an HII region around a B0 ZAMS star.

The protostellar interpretation requires mass accretion rates higher than the typical values for the formation of low- and intermediate-mass stars discussed in earlier chapters. We have seen, in fact, that at a rate of 10^{-5} $M_\odot$ yr^{-1}, a protostar joins the main sequence at a mass of ~8 $M_\odot$ where it releases ~3000 $L_\odot$, less than the observed values in excess of 10^4 $L_\odot$. Thus, the need for higher accretion rates, also suggested by the large mass *loss* rates inferred from the associated outflows (see [84]).

We have reached an important conclusion: in order to distinguish between the two plausible explanations, massive protostars vs. young ZAMS stars, the critical quantity to determine observationally is the mass accretion rate. Future high angular resolution millimeter observations will reveal whether the selected sources are characterized by *high* values of the accretion rate (protostars) or not (ZAMS stars). Given the large distances of massive star forming regions ($\gtrsim$1 kpc), one has to wait for the completion of (sub)mm-wave arrays such as ALMA to reach the desired spatial resolutions.

11.2 The Evolution of Ultracompact HII Regions

The presence of an UC HII region is the earliest manifestation of massive stars of spectral type earlier than B3. Observationally, UC HII regions are small ($r \lesssim 10^{17}$ cm), dense ($n \gtrsim 10^4$ cm^{-3}), high pressure ($nT \gtrsim 10^8$ cm^{-3}K) HII regions found in about 10% of the O stars in the Galaxy (cf. [105]). This frequency of occurrence is quite high and leads to the following paradox. The lifetime of an UC HII region is approximately given by

$$\mathrm{t_{UC}} \sim 0.1\,\mathrm{t_{O6}} \sim 3 \times 10^5\,\mathrm{yr}\,, \tag{81}$$

where $\mathrm{t_{O6}} = 3 \times 10^6$ yr represents the main-sequence lifetime of an O star. The dynamical time is

$$\mathrm{t_{dyn}} = \frac{r}{a_T} \sim 3 \times 10^3\ \mathrm{yr}\,, \tag{82}$$

where a_T is the sound speed, here taken to be ~10 km s^{-1}. Thus, the expansion of the UC HII region should take place very quickly and very few of them should be observed at any given time.

Several explanations have been proposed to explain the long duration of the ultracompact phase. However, before discussing models in some detail, let us first see the basic fact that the UC regions cannot be classical HII regions. The main reason is that the observed radii are much smaller than the typical size of a developed HII region. To see why, consider the radius of the Strömgren sphere

$$r_S^{\mathrm{in}} = \left(\frac{3\phi}{4\,\pi\, n_\circ^2\, \beta_2}\right)^{1/3} \sim 0.02\,pc\,\phi_{48}^{1/3}\, n_5^{-2/3}\,, \tag{83}$$

where ϕ is the stellar ionizing photon rate (s^{-1}), $\phi_{48} \equiv \phi/10^{48}$ s^{-1}, β_2 is the hydrogen recombination rate coefficient of levels $n \gtrsim 2$, $n_\circ$ is the number density and $n_5 \equiv n_\circ/10^5$ cm^{-3}. The evolution of the Strömgren radius with time is given by

$$r_S = r_s^{\mathrm{in}} \left(1 + \frac{7}{4}\frac{a_T\, t}{r_s^{\mathrm{in}}}\right)^{4/7}\,. \tag{84}$$

If we take as a lower limit to the lifetime of the ultracompact region $t = 10^5$ yr, then eq. (84) indicates that $r_S \sim 10^{18}$ cm~0.3 pc, a factor of at least 5 larger than the observed size of an UC HII region.

One possibility to solve the problem of the lifetimes is that the *ram pressure* from the infalling gas provides pressure to confine the gas. Although very interesting, and initially supported by the observations of OH lines which seemed to indicate collapse ([80]), this explanation however fails since the balance between ram and thermal pressure is unstable to small perturbations ([44]). A simple argument can show why this is true. The thermal pressure is given by $P_{\mathrm{th}} \sim n_e T_e$. Now, the ionizing flux is approximately

$\phi \sim n_e^2 r^3$; since ϕ ~const, then $n_e \sim r_s^{-3/2}$. The electron temperature is $T_e \sim 10^4 K \sim const$ and finally

$$P_{\rm th} \propto n_e T_e \sim r^{-3/2}\,. \tag{85}$$

On the other hand, the dependence of the ram pressure of the infalling gas on the density can be estimated in the following way. The ram pressure is equal to $P_r = \rho V_{\rm inf}^2$, where $V_{\rm inf}$ is the infall velocity. From the equation of continuity, the mass accretion rate is given by $\dot{M} = 4\pi r^2 n m_H V_{\rm inf}$. Since $V_{\rm inf} \propto r^{-1/2}$ and assuming $\dot{M}$ =const, it follows that $n \propto r^{-3/2}$. Thus, the ram pressure is

$$P_{\rm r} \propto n\, V_{\rm inf}^2 \sim r^{-5/2}\,. \tag{86}$$

Now, it can be readily seen why a small perturbation will make the equilibrium unstable. In fact, a displacement to larger r will produce a larger decrease in $P_{\rm r}$ than in $P_{\rm th}$ and the HII region will expand freely. In the opposite limit, the ram pressure can overcome the thermal pressure and the infalling matter will crash the HII region. The equilibrium condition corresponds to the derivation of the critical mass accretion rate for choking-off the radio emission from a massive protostar.

More realistic models have been proposed to solve the lifetime problem. Among them, the possibility that the UC HII region is formed in bow shocks caused by the stellar motion through dense molecular gas ([96]). In this model, the equilibrium is established between the pressure of the stellar wind coming from the O star and the ram pressure of the infalling gas. Unlike the previous case, such a balance is stable. This model successfully accounts for the cometary UC HII regions which represent about 20% of the Wood & Churchwell sample. However, a major limitation is that the required stellar velocities ($v_* \sim$5-10 km s^{-1}) appear uncomfortably high to explain *all* the observed ultracompact HII regions, considering the typical velocity dispersions of only few km s^{-1} observed in young stars.

Finally, an interesting mechanism based on the idea of photoevaporation of accretion disks surrounding O stars has been suggested by Hollenbach et al. [44] Here, the diffuse radiation field due to hydrogen recombination in the photoionized gas produces a steady evaporation at the surface of the disk, producing a *disk wind*. In this case, the lifetime of the UC HII region depends on the mass of the disk. Using typical parameters of disks, $r_d \gtrsim 10^{15}$–10^{16} cm and $M_d \sim 0.3 M_* \sim$3–10 M$_\odot$, the characteristic lifetime of the photoevaporating disk is

$$t_d = \frac{M_d}{\dot{M}_*} \sim 7 \times 10^4\, yr\, \phi_{49}^{-1/2} \left(\frac{M_*}{10 M_\odot}\right)\,. \tag{87}$$

Then, if $M_d \sim$2–10 M$_\odot$, t$_{\rm UC} \gtrsim 10^5$ yr, in agreement with the observational constraints. But are these rather big disks commonly observed around massive stars? We will come back to this important issue later on in the chapter.

12 Clustering of Young Stars

Direct and indirect observations suggest that most, if not all, stars are formed in groups rather than in isolation ([16]). An important result that strongly constrains theories of massive stars and stellar clusters formation is that the stellar density of young stellar clusters seems to depend on the most massive star in the cluster. Low-mass stars are usually found to form in loose groups with typical densities of a few stars per cubic parsec ([33]), while high-mass stars are found in dense clusters of up to 10^4 stars per cubic parsec (e.g. [42]). Therefore, the transition between these two modes of formation should occur in the intermediate-mass regime, namely for $2 \lesssim M/M_\odot \lesssim 15$.

Recently, an extensive near infrared (NIR) survey for young clusters around 44 intermediate mass stars has been completed with the goal of detecting and characterizing the clustering properties of intermediate-mass stars ([93]). At NIR wavelengths, and especially at K-band (2.2 μm), the low extinction allows the detection of embedded young stars in the vicinity of the bright Herbig star. Figures 46 and 47 show two fields surrounding the isolated star IP Per (spectral type A3e) and a rich cluster associated with BD+65°1637 (B2e). Given the distance and expected ages of these two targets, the completeness absolute magnitudes at K-band correspond to a minimum detectable stellar mass of less than 0.1 $M_\odot$ (see Sect. 6.3).

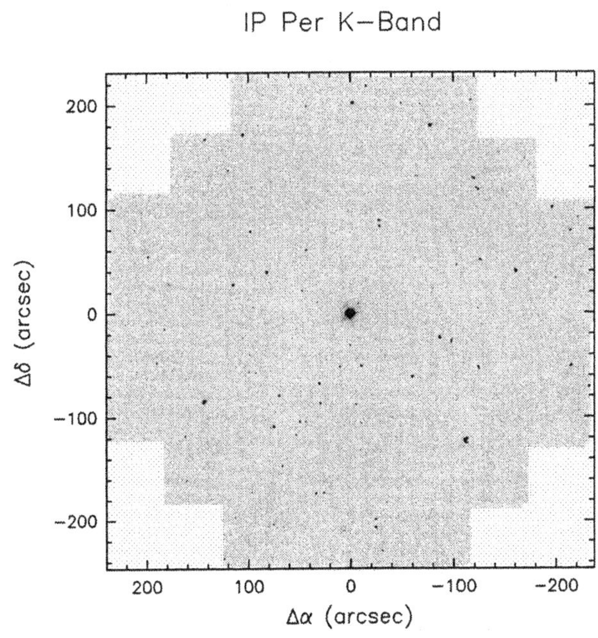

Fig. 46. K-band image of the isolated Herbig star IP Per (spectral type A3). (From Testi et al. [93])

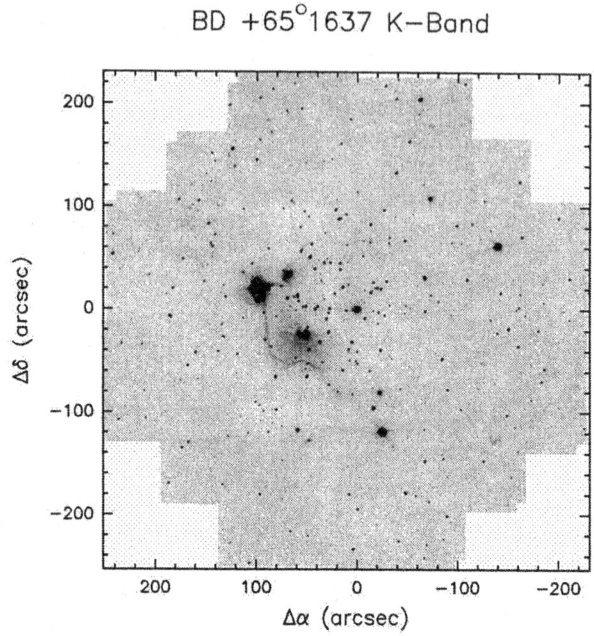

Fig. 47. K-band image of the cluster associated with BD+65°1637 (spectral type B2). (From Testi et al. [93])

The most reliable *richness indicator* of young clusters is the integral of the radial surface stellar density profile, I_C, defined as:

$$\mathrm{I_C} = \sum_{i=0}^{i_{max}} \pi(n(r_i) - n_\infty)(r_{i+1}^2 - r_i^2), \tag{88}$$

where $n(r_i)$ is the local source surface density (i.e., the number of stars between r_i and r_{i+1} arcsec from the Herbig Ae/Be star divided by the area $\pi(r_{i+1}^2 - r_i^2)$. $imax$ is chosen so as to contain all the members of the clusters and the field stars surface density (n_∞) is estimated at the edge of each field. In all fields $\Delta r \equiv r_{i+1} - r_i = 12''$, which ensures a good resolution and (in most cases) a reasonable number of stars per annulus.

The quantity I_C thus estimates the number stars above the background. Stars associated with rich clusters have a value of $I_C \gtrsim 40$. In the intermediate regime with $10 \lesssim I_C \lesssim 40$, a cluster may be present. However, when $I_C \lesssim 10$ only small aggregates or background stars are found, a situation similar to that found around T Tauri stars.

An important result of the observations is that all the detected clusters appear to have similar sizes, with a typical radius of ~0.2 pc (FWHM), a value which which is remarkably similar to the dimensions of dense cores in molecular clouds. Assuming spherical clusters, we can use this value to convert I_C into stellar volume densities. The result is illustrated in Fig. 48 where the two solid vertical lines represent the typical ranges of densities for low mass stars aggregates in Taurus–Auriga and for the Orion Nebula Cluster. Note how stars later than B9 are not surrounded by dense (> 150 stars/pc^3) stellar groups, whereas only the early type B0-B2 may be associated with stellar clusters with densities exceeding 10^3 stars/pc^3. The observed trend indicates that there is a physical relationship between cluster density and the maximum stellar mass (Hillenbrand [41]; Testi et al. [94]). The intermediate mass-stars with spectral types between A5 and B0 mark the transition between the low density aggregates of $\lesssim$10 stars pc^{-3} typical of T Tauri stars and the dense clusters.

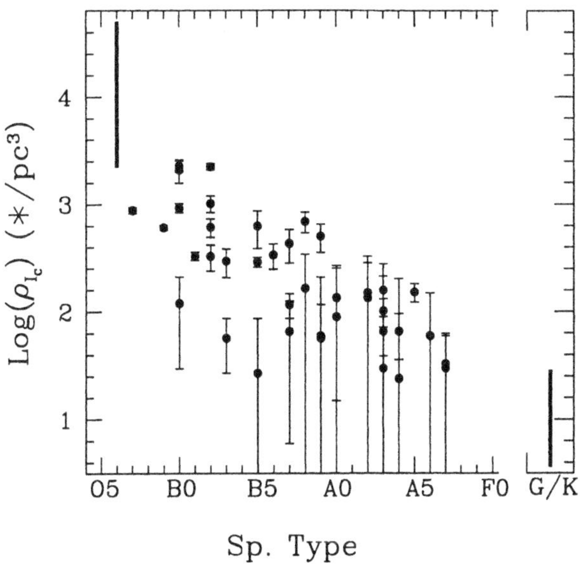

Fig. 48. Mean stellar density of the stellar groups associated with Herbig Ae/Be stars of a given spectral type. The two solid vertical lines represent the typical ranges of densities for low mass stars aggregates in Taurus–Auriga and for the Orion Nebula Cluster. (From Testi et al. [93])

The fact that more massive stars are typically found in *denser* and *not* simply richer groups suggests that this is a necessary condition for their formation. Then, the formation of massive stars requires special physical conditions that can only be obtained in the early evolution of dense clusters and it is not the result of random sampling in large clouds ([19]). These results tend to support the idea that coalescence in dense clusters may play a rôle in the formation of the most massive stars (see [16]; [87]).

12.1 Clustering and Mass Segregation

Observational studies of stellar clusters find a strong tendency toward *mass segregation*. That is, the average stellar mass increases toward the system's center, which is also the densest region. This is the case, for example, of the Pleaides where all of the brightest stars are concentrated in an inner core which covers only a small fraction of the cluster radius of ~10 pc (Raboud & Mermilliod [79]). The high concentration of the more massive objects in the center of the cluster is evident in Fig. 49 that shows the position of 270 members brighter than $V = 12.0$. Mass segregation, however, is observed even in the youngest clusters, such as Mon R2, NGC 2024 and the Trapezium cluster where the effect is seen down to stars of a few solar masses (Carpenter et al. [12]; Hillenbrand & Hartmann [42]).

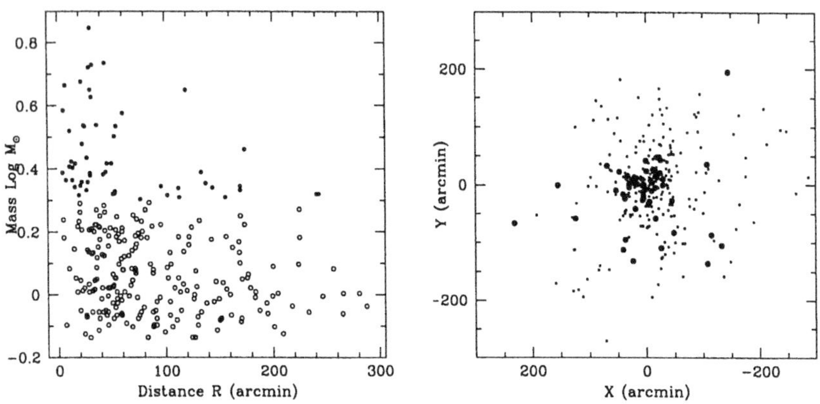

Fig. 49. Mass segregation in the Pleiades cluster for 270 members. The *left panel* shows the stellar mass as a function of the radial distance from the cluster center. The spatial map of the same stars is displayed in the *left panel*, where the size of the dots is related to the mass of the star, with bigger symbols representing more massive objects. (From Raboud & Mermilliod [79])

It is well known that gravitational interaction between individual stars transfers kinetic energy to the lowest mass stars and drives the system toward energy equipartition. Thus, massive stars tend to sink toward the center of the cluster, while those of lesser mass are ejected on wider orbits. However, the main problem with this mechanism is the time scale. The transfer of energy requires many interactions and occurs on the *relaxation time*:

$$t_{\rm relax} \sim \frac{N}{8\,lnN}\, t_{\rm cross}\,, \tag{89}$$

where N is the number of stars in the cluster, and t_{cross} is the dynamical time scale, or *crossing time*

$$t_{\mathrm{cross}} = \frac{2\,R_{1/2}}{\sigma}\,, \tag{90}$$

where $R_{1/2}$ is the half mass radius and σ the one dimensional dispersion velocity. Now, for the Pleiades, $N \sim 1000$, $\sigma \sim 0.5$ km s^{-1}, and $R_{1/2} \sim 2$ pc. Then, $t_{\mathrm{cross}} \sim 4 \times 10^6$ yr, and $t_{\mathrm{relax}} \sim 18\ t_{\mathrm{cross}}$. Since the cluster age is about 10^8 yr, dynamical relaxation has had time to operate effectively. On the other hand, for the Trapezium cluster, $t_{\mathrm{relax}} \sim 25\ t_{\mathrm{cross}}$, whereas its estimated age is only about 1.5×10^6 yr, or $\sim 3\ t_{\mathrm{cross}}$ (Palla & Stahler [42]). Thus, the massive stars of the Trapezium cluster did not have the time to migrate to the center from larger distance as a result of mutual interactions. Indeed, the results of N-body simulations of cluster evolution indicate that clusters that are not already segregated at birth cannot relax rapidly enough to reproduce the observed distributions (Bonnell & Davies [9]).

On the other hand, Kroupa [50] has followed the stellar dynamical evolution of bound clusters, considering a high percentage of primordial binaries. Contrary to the other results, Kroupa finds that massive stars do sink to the cluster center on a time scale given approximately by $(< M_* > /M_{\mathrm{massive}})t_{\mathrm{relax}} \lesssim 0.1 t_{\mathrm{relax}}$. This process leads to cluster heating and expansion in the first few million years. Also, because of mutual interactions in the high density core, massive stars can be readily ejected from the cluster with high velocities. Measurements of the space velocity of massive stars away from clusters/associations are required to verify the model predictions. Since purely stellar dynamical models do not account for the presence of the molecular gas within the cluster, the finding of a rapid onset of mass segregation should be considered as preliminary. In conclusion, the primary question is *not* how massive stars find their way to the densest, inner region of a cluster. The problem is rather why they preferentially *form* in that location.

12.2 Effects on the Environment

An important, and still open, question is whether intermediate- and high-mass stars are surrounded by circumstellar disks. While disks of gas and dust are known to be associated with the majority of classical T Tauri stars, the presence of similar structures around more massive objects is still elusive. Clearly, if Herbig stars do not have (and never had) circumstellar disks , the natural conclusion is that their formation mechanism differs qualitatively from that of lower mass stars which requires the presence of an accretion disk.

Natta et al. [66] have reviewed the status of the observational evidence on Herbig stars (the so-called "disk debate"). The analysis of interferometric data indicates that there is a clear difference between the frequency of occurrence of disks around HBe ($M_* \gtrsim 5\ M_\odot$) and HAe ($M_* \lesssim 5\ M_\odot$) stars: for the latter, the frequency is very large (between 75% and 100%), whereas for the more massive stars the incidence of disks drops to almost zero. Interestingly, it appears that while the disk mass seems to decrease with the stellar mass, the ratio of disk mass to stellar mass is roughly constant for stars in the range A0–M7. Thus, it appears that PMS stars of all masses have relatively low-mass disks ($M_{disk}/M_* \sim 0.04$). The available measurements in both T Tauri and Herbig stars are displayed in Fig. 50.

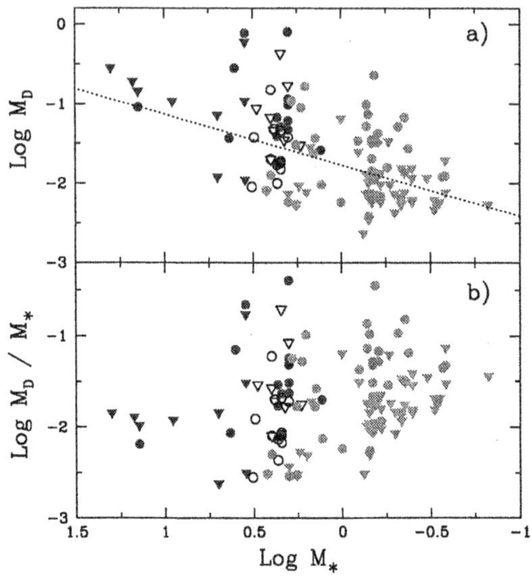

Fig. 50. Properties of disks around low-mass (T Tauri) and intermediate-mass (Herbig Ae/Be) stars. Panel a): Total disk mass (gas+dust) vs. the stellar mass. Panel b): ratio of the disk mass over the stellar mass as a function of M_*. Detections are shown by circles, 3σ upper limits by triangles. Filled symbols refer to stars with single dish observations, open symbols to interferometric measurements. (From Natta et al. [66])

The absence of evidence of disks around early-type stars is not evidence of genetic absence. A possible explanation may be due to the rapid dissipation of the circumstellar material, mainly as a result of the effect of the intense radiation field from the central stars. Maps at far-infrared and millimeter wavelengths show that often the Herbig Be stars are located in large cavities: Fuente et al. [27] find a time scale of less than 10^6 yr for a complete dispersal of the surrounding dense gas and dust. An important result is that the efficiency of the dispersal is very different for early-type (B0–B5) and late-type (B5–A5) stars. While in early-type stars the mean gas density in a radius of 0.08 pc decreases by almost two orders of magnitudes during their evolution to the main sequence, in late-type objects it decreases by less than an order of magnitude. The morphology of the molecular cloud around several early-type Be stars is shown in Fig. 51. The fact that Herbig Be stars are associated with rich clusters of lower mass stars (see below) can also partially account for the rapid evolution of the environment and the lack of circumstellar disks. However, the stellar density in these clusters does not seem to be high enough to induce significant star-disk interactions and dissipation.

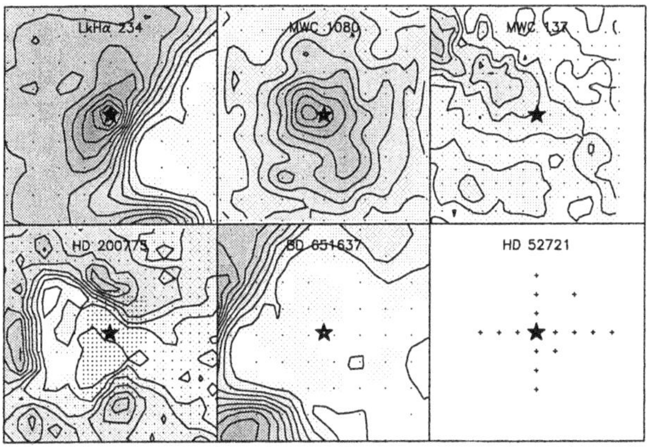

Fig. 51. Gas dispersal around Herbig Be stars. The maps show the integrated intensity emission of the ^{13}CO J= 1 → 0 line from stars of spectral type B0 to B5. The size of the box is 1pc×1pc for all the objects. Contour levels are 0.3 to 70 by 4 K km s^{-1}. The star position is marked with an asterisk. Note how the youngest objects LkHα 234, MWC 1080) are still embedded in the dense gas, whereas the more evolved ones sit in the middle of an large cavity. (From Fuente et al. [27])

The evidence for the presence of disks around massive stars is also as uncertain as in the case of the Herbig Be stars. Only recently, sensitive, high resolution observations of dust and molecular gas emission have become available for a small sample of objects. The distribution of SiO masers around the deeply embedded source IRc2 in Orion has been interpreted as due to the presence of a rotating and expanding circumstellar disk, about 80 AU in diameter (Plambeck et al. [76]). One of the most convincing cases is represented by the source IRAS 20126+4104 with a luminosity of $\sim 1.5 \times 10^4$ L$_\odot$. Observations in CH_3CN have revealed the existence of a flattened disk-like structure with a radius of $\sim 5 \times 10^3$AU and with a velocity gradient across the major axis (Cesaroni et al. [14]; Zhang et al. [108]). The kinematics of the disk are consistent with Keplerian rotation around a 20 M$_\odot$ star. The small scale structure of this object is shown in Fig. 52.

Interferometric radio observations of masers, both H_2O and CH_3OH masers, have indicated that the spots often lie across the diameter of the associated

UC HII regions, suggesting that they are tracing disklike structures (Norris et al. [71]). The derived properties of these disks are rather extreme: in the case of the methanol masers, the diameters are of the order of $3-4\times10^3$ AU, surrounding central objects of masses in excess of ~10 $M_\odot$. Further support to the disk hypothesis has been provided by by mid-infrared imaging, as in the case of G339.88−1.26, which has revealed polarization patterns typical of the scattering lobes in bipolar cavities (Stecklum et al. [88]). However, the complex morphology observed in these regions makes the unambiguous identification of disks still uncertain, since the infrared features can also be attributed to scattering peaks while the central structures remain undetected because of the very large extinction ($A_V \gtrsim 100$ mags). Thus, the rôle and relevance of disks in the earliest phases of the evolution of massive stars are still uncertain, but the high angular resolution in the thermal infrared offered by the large telescopes now available holds great promises to directly detect these elusive structures.

12.3 Clustering and Binary Formation

We have shown that young stars form in groups/clusters. But does clustering extend to all spatial scales, or, is there a characteristic threshold? In an interesting paper, Larson [54] has extended the analysis of the clustering of T Tauri stars by Gomez et al. to even smaller separations by using the rich database available on binary/multiple stars in Taurus-Auriga . Larson finds that binary stars exhibit self-similar clustering on the largest scales, but that there is a definite break from self-similarity at a scale of about 0.04 pc. This change separates the regime of binary and multiple systems on smaller scales from that of true clustering on larger scales. At $\theta < 0.04$ pc, the power-law corresponds to a distribution of binary separations between 10 and 8000 AU. The break in the observed distribution is interpreted as evidence for an intrinsic scale in the star formation process which Larson identifies with the Jeans length in a typical cloud core. However, as we have seen in eq. (14), $\lambda_J \sim 0.2$ pc, a factor of at least five bigger than the break-up scale. Finally, the observed hierarchical clustering can be interpreted as a fractal structure with a dimension of about 1.4, possibly originating from pre-existing hierarchical structures in the progenitor molecular clouds (e.g., Elmegreen & Falgarone [22]).

Additional support to the findings in Taurus-Auriga has come from recent studies of the binary population in Ophiucus and in the Trapezium cluster (Simon [85]), and in a number of other star forming regions (Nakajima et al. [64]; Gladwin et al. [32]). The situation is summarized in Fig. 53. The surface density distribution $\Sigma(\theta)$ is characterized by a binary regime in which $\Sigma(\theta) \sim \theta^{-2}$, and a large scale clustering where $\Sigma(\theta) \sim \theta^{-0.5}$: Although the distributions are rather similar, the characteristic scale of the break differs: from ~400 AU for the Trapezium to ~12000 AU for Taurus, or even higher (~20000 AU) in Chamaeleon. The interpretation is that, in addition to the

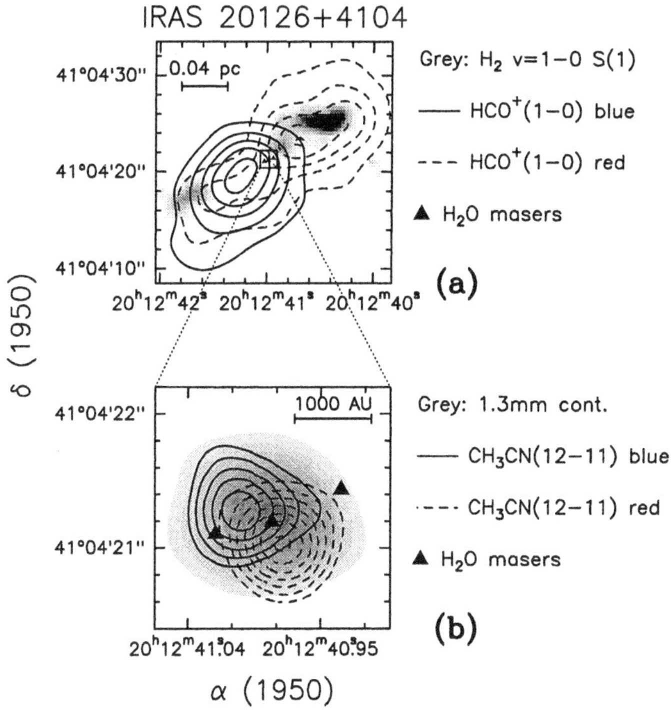

Fig. 52. Disk around the massive source IRAS 20126+4104. Top panel: overlay of the H_2 2.1 μm line emission (grey scale) and HCO^+ (1–0) emission, showing the presence of a bipolar outflow. Bottom panel: enlargement of the central region with the grey scale representing the 1.3 mm continuum emission. The full and dashed lines show the CH_3CN (12–11) emission integrated under the blue and red line wings, showing evidence for Keplerian rotation. Finally, the position of the H_2O masers is indicated by the triangles. (From Kurtz et al. [51])

Jeans length, another parameter, the stellar density, should operate to establish the transition from binary companions to clusters. However, Bate et al. [7] have carefully reanalyzed the data base used by these groups and reached a somewhat different conclusion: the observed power law slope of the mean surface density of companions is also consistent with clustering on a single scale, with no evidence for a fractal distribution.

12.4 Clustering and Star Formation Efficiency

The second property of clustering bears directly on the star formation mechanisms in dense regions. It is clear that the standard model of isolated star formation mediated by the action of ambipolar diffusion is hardly tenable in these conditions. Therefore, other mechanisms must be invoked. It is well known that, although most star formation occurs in molecular cores that

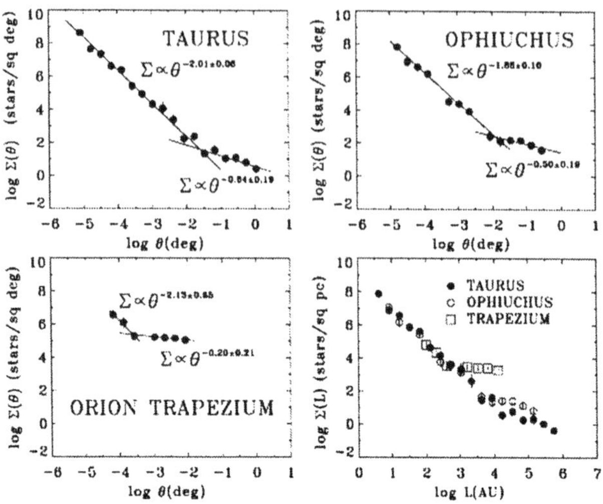

Fig. 53. Variation of the stellar surface density of companions with angular separation in different star forming regions. The right hand bottom panel shows the results for the three regions placed on a common scale of apparent separations in AU. (From Simon 1997)

form large groups of stars, most main sequence stars are *not* found in clusters. Thus, most of the embedded clusters that have been found in star forming regions cannot survive for long as bound open clusters and the majority must be dispersed soon after their formation. The mosty likely physical reason for this behavior is the low efficiency of star formation in clusters that do not remain gravitationally bound after gas removal. Lada et al. [53] have shown that the formation of a bound cluster requires at least a 30% efficiency of conversion of gas into stars. This value is much higher than the typical 2-3% estimated in molecular clouds, but it also higher than that derived for the young cluster in ρ Ophiuchi (about 20%). In the case of the Trapezium cluster, the situation is more favorable since the total mass of stars is comparable to the mass of gas in the region, so that the efficiency there (~50%) seems to have been high enough to produce a bound cluster.

Models of the early evolution of young clusters still embedded in the parent gas are needed to firm up the estimate of the minimum star formation efficiency needed to form bound clusters. The important result of Lada et al. is based on a simplified dynamical model in which the gas was assumed to be lost to the system with an ad hoc time-law prescription. Recent improvements using N-body+ fully hydrodynamical codes allow to follow the radiative and mechanical interaction between stars and gas in time and therefore the mechanisms by which the binding energy is removed from the system. For example, it is found that most of the mass is lost via the action of stellar winds and outflows during the earliest phases of the evolution, before the most massive stars evolve to supernovae. Also, there is a strong dependence on the stellar mass distribution and on mass segregation. In fact, there is evidence that massive stars are born at the center of the clusters and do not settle there via dynamical friction and gravitational drag. The formation of the most massive objects can thus occur only at the center of dense clusters, where the deep potential well of the surrounding gas and interaction with close objects may allow accretion to continue onto very luminous stars.

12.5 Clustering and the IMF

This point introduces the last question of what determines the mass of a star in a cluster. We have seen that in all YSOs infall and outflow occur simultaneously. Also, the amount of mass which is available from cloud cores is generally bigger than the final stellar mass. Thus, it is not because of lack of reservoir that stars stop accreting. It is quite natural then to identify in the outflows the main agents for reversing the infall and limiting the final stellar mass. This idea has been put forward initially by Norman & Silk [70] and more recently by Adams & Fatuzzo [1] and Nakano et al. [65]). The main idea is that the mass outflow pushes out the surrounding core material, producing a bubble and a thin dense shell. The condition for stopping infall is achieved when the shell radius $R_{\rm sh}$ grows to the core radius $R_{\rm c}$, or

$$R_{\rm sh} \sim \left(\frac{\dot{M}_{\rm out} V_{\rm out}}{\rho_{\rm c}}\right)^{1/4} t^{1/2} \approx R_{\rm c} \sim \left(\frac{M_{\rm c}}{\rho_{\rm c}}\right)^{1/3} , \tag{91}$$

where $\dot{M}_{\rm out}$ and $V_{\rm out}$ are the mass outflow rate and outflow velocity, respectively. Using typical values for the outflow and core properties, Nakano et al. obtain a value of the typical mass of about 2 $M_\odot$ with a dependence on the mass infall and outflow rates $M_* \sim (\dot{M}_{\rm out}/\dot{M}_{\rm inf})^{7/6}$. The estimated star formation efficiency is found to be less than ~5%, in agreement with the observations.

The main problem with the scenario in which the mass of a star is set by the outflow mass rate is that the condition in eq. (91) is achieved when the bubble has reached a value of ~0.2-0.4 pc, i.e. the core radius. Now in the conditions of dense clusters, the typical separation varies between 0.04 pc (Trapezium cluster) and 0.12 pc (dense cores in L1641). Both values are smaller than the shell and core radii. Thus, accretion is shut off not because of outflows, but because of lack of available material and by interaction with neighboring protostars. In other words, the small extent of the protostellar envelopes does not allow the outflow to propagate to large distances and therefore would limit the total amount of matter that can be accreted to much smaller values than the characteristic solar mass.

As a possible alternative, it is interesting to envisage a model in which the stellar mass is not set by the feedback effects of forming stars on their environment, but by the clumpy structure of molecular clouds. Strong evidence exists for clumpy substructures down to the smallest scales. Also, the clump mass spectrum is a power-law $N(M_{\rm cl}) \sim {\rm M}_{\rm cl}^{-1.6}$, for $\rm M_{cl} \gtrsim 1~M_{\odot}$. Thus, it is natural to simply relate the clump mass spectrum to the stellar mass. Indeed, the stellar IMF is a power-law $N(M_*) \sim M_*^{-1.7}$ at subsolar masses and becomes much steeper at larger masses ($\sim M_*^{-2.3}$). The similarity of the two power indices suggests a linear relation between the mass of the clump and that of the star, with an efficiency of conversion of gas into star of about 100%. At higher masses, the efficiency decreases rapidly, probably due to subfragmentation of more massive clumps, implying that it becomes more and more difficult to build up massive stars. In order to reconcile this scenario with the observed stellar IMF, the star formation efficiency should scale with the mass of the clump as $M_{\rm cl}^{-0.7}$. This can be achieved if massive stars result from the coagulation of several protostellar cores in the same clump. Interestingly, some support to the possibility that the IMF in clusters is basically determined by the clump mass distribution has been presented by Motte et al. [63] and Testi & Sargent [92] from the mass spectra of pre-stellar clumps mapped in the ρ Ophiuchi and Serpens molecular clouds. The derived mass distribution in ρ Ophiuchi is displayed in Fig. 54. It appears that the mass distribution follows approximately $\Delta N/\Delta M_{\rm cl} \propto M_{\rm cl}^{-1.5}$ below $M_{\rm cl} \sim 0.5~\rm M_{\odot}$, then it steepens at larger masses. This break in the power law is reminiscent of that observed in the stellar mass spectrum.

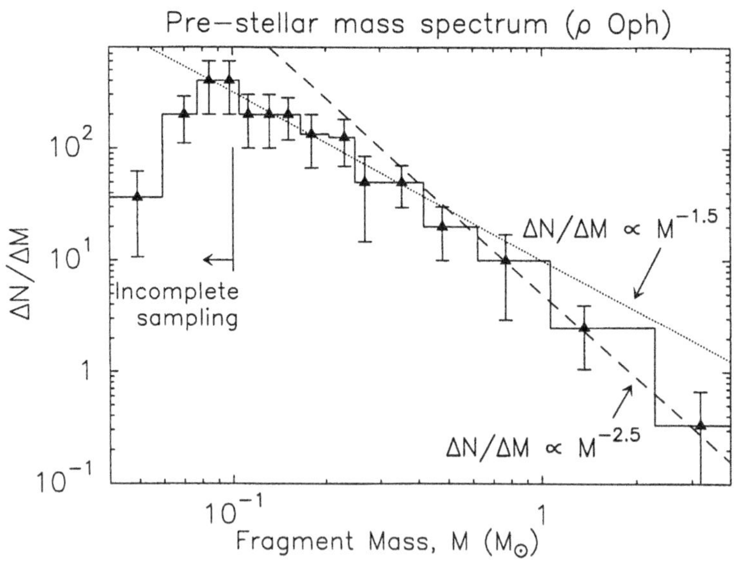

Fig. 54. Mass spectrum of 59 pre-stellar clumps in ρ Ophiuchi. The dotted and dashed lines show two power laws $\Delta N/\Delta M_{\rm cl} \propto M_{\rm cl}^{-1.5}$ and $\Delta N/\Delta M_{\rm cl} \propto M_{\rm cl}^{-2.5}$, respectively. (From André et al. [1])

13 Stellar Populations in Clusters and Associations

The previous chapter has emphasized the rôle of young stellar clusters for testing star formation theories. One important clue is the actual pattern of stellar births within any region. The most powerful technique in this regard is to place the stars in the H-R diagram: one can then read off directly the ages and masses of individual cluster members and reconstruct the detailed history of star formation, as well as the relative production rates of various masses. Studies of the stellar population of young clusters have a venerable tradition, dating back to the original works of Walker [97] and Herbig [37].

Herbig was the first to argue that low- and intermediate-mass stars form first in clusters and that the formation of one or more O stars may terminate the process by dispersing the ambient gas. The spread in the formation times of PMS stars was then quantified by Iben & Talbot [47] who used theoretical tracks and isochrones in the H-R diagram to reconstruct the properties of the young clusters NGC 2264 and NGC 6530. They established that a typical group is created over an interval of some 10^7 yr, and confirmed Herbig's suggestion on the progressive formation of more massive objects with time.

Many subsequent studies have used the same basic tools to investigate several other clusters, and the picture emerged of "sequential star formation", in which the average stellar mass being produced at any time increases in a smooth fashion (e.g., Herbst & Miller [40]). Building upon the results of protostellar evolution, Stahler [86] examined critically the notion of a correlation between stellar masses and ages and concluded that the data available at the time were more consistent with a scenario in which all stellar masses form approximately in the same interval of time within a given cluster. Thus, the idea of "contemporaneous star formation", with a probability based on the initial mass function (Elmegreen & Clemens [21]).

The improvement over the last decade in the empirical knowledge of the stellar population of nearby star forming regions, and the simultaneous progress in the theory of PMS evolution call for a modern reassessment of these basic issues. With such an approach, one hopes to address not only the problem of the star formation history of individual clusters and associations, but also what factors determine whether the resulting groups become aggregates of T Tauri stars, bound clusters, or expanding OB associations. Finally, a solution can be found to the long standing puzzle of the large deficit of T Tauri stars of age greater than about 10^7 years, the so-called post-T Tauri problem (Herbig [31]). In this chapter, we will provide first a brief description of the empirical properties of the stellar populations in star forming regions in the solar vicinity. The characterization of the age distributions rests on the evolutionary tracks and isochrones described earlier, and will be presented separately. The common evidence is for a pronounced *acceleration* in star formation activity that lasts for several million years in any given region. This pattern represents a true physical phenomenon and may arise from a global contraction of the parent molecular cloud. The implications on the post-T Tauri problem will be addressed at the end.

13.1 H-R Diagrams

Taurus-Auriga Because of its proximity and overall low extinction ($d =$ 140 pc, $A_V \sim 1$), the Taurus-Auriga molecular complex has been extensively surveyed for PMS stars at all wavelengths. As a result, the current census indicates more than 150 known members, most of them concentrated within several distinct clumps of molecular gas. Figure 55 shows the distribution of this population, relative to the molecular gas as outlined by the ^{12}CO line emission. The optical and near-infrared properties have been presented by Kenyon & Hartmann [52] who determined luminosity and effective temperatures for about 105 stars, including 65 CTTS and 38 WTTS. This sample of optically visible PMS stars is believed to be nearly complete for objects with V-magnitude less than about 15, corresponding to stars of about 0.3 $M_\odot$. The presence of 25 embedded Class I objects in Taurus-Auriga indicates that the population of the youngest sources may be incomplete. Kenyon et al. [48] have characterized the properties of six of them, bringing the tally to 108 stars. Only two class 0 objects (L1527 and IRAM 04191) have been identified to date.

The resulting distribution of stellar population in the H-R diagram is shown in Fig. 56. The vast majority of the stars lie between the birthline and the ZAMS. Most stars stars have masses between $\sim$0.1 and 2 $M_\odot$ and their contraction ages vary between 10^5 and 10^7 yr, with few WTTS stars close to or slightly below the ZAMS. The median mass of the sample is about 0.6 $M_\odot$, with a relative dearth of stars with masses below $\sim$0.2 $M_\odot$, possibly due to selection effects. However, both deep optical surveys of individual dark clouds (Briceño et al. [17]; Luhman & Rieke [60]) and large scale, moderately deep optical and X-ray surveys of the molecular cloud complex (Briceño et al. [11]) have failed to detect a large population of very low mass, faint and more evolved objects.

The completeness issue should also consider the information provided by the ROSAT All-Sky-Survey (RASS). Wichmann et al. [101] have identified 76 new T Tauri star candidates over a large area in the direction of the Taurus-Auriga region. Most of them (72) were classified as WTTS on the basis of weak Hα emission or absorption and detection of lithium in low-resolution spectra. The high number of WTTS identified would yield a WTTS/CTTS ratio of about 6, a value close to that predicted by Walter et al. [98] on the basis of the EINSTEIN observations of a much more limited area. However, observations of the lithium line at high spectral resolution of a subsample of the WTTS has reduced the number of bona-fide young objects, the rest being more evolved objects (Martín & Magazzù [59]). However, without a precise estimate of the distance of the RASS stars at large distances from the star

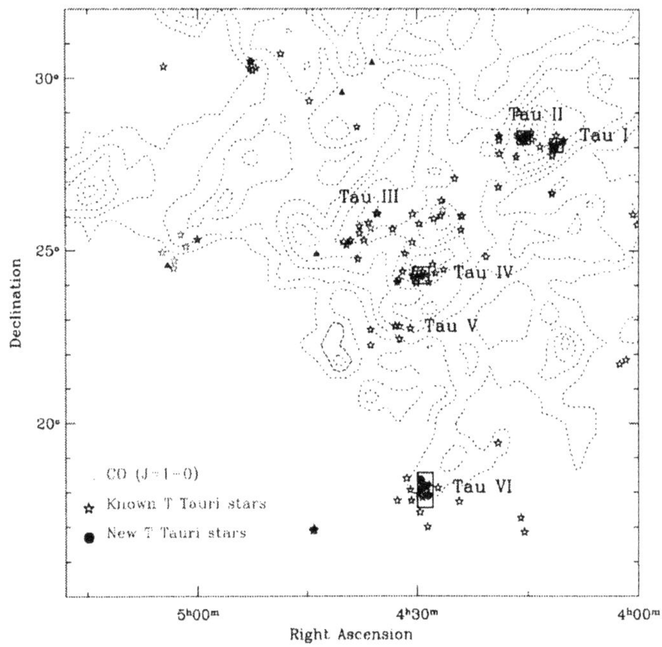

Fig. 55. Location of known PMS stars in Taurus-Auriga. The six main stellar groupings identified by Gomez et al. [33] are indicated. The dashed contours trace the extent of the ^{12}CO (J=1–0) line emission, obtained by Ungerechts & Thaddeus [95]. (From Briceño et al. [17])

forming region, it is impossible to place them in the H-R diagram and derive masses and ages.

Lupus The Lupus region, at a distance d =140 pc, contains four T associations (Lupus 1 to 4) associated with small dark clouds , embedded in a larger CO complex of mass comparable to Taurus-Auriga. Most of the star formation activity is concentrated in Lupus 3, where Schwartz [81] identified 69 Hα emission line stars. With the exception of two Herbig Ae stars (HR 5999 and HR 6000), all the other members are of spectral type later than M0 (Hughes et al. [46]). The ROSAT survey has discovered some 130 new stars, distributed over an area of more than 200 square degrees (Krautter et al. [49]). Follow-up observations have established the PMS nature for 27 of them, located within the boundaries of CO emission (Wichmann et al. [102]). Thus, the ratio of WTTS to CTTS appear to be about unity. The H-R diagram for the population of 92 stars is shown in Fig. 56. Most stars have masses below ~0.6 $M_{\odot}$, and are well distributed around and below the birthline. The Lupus H-R diagram resembles that of Taurus-Auriga.

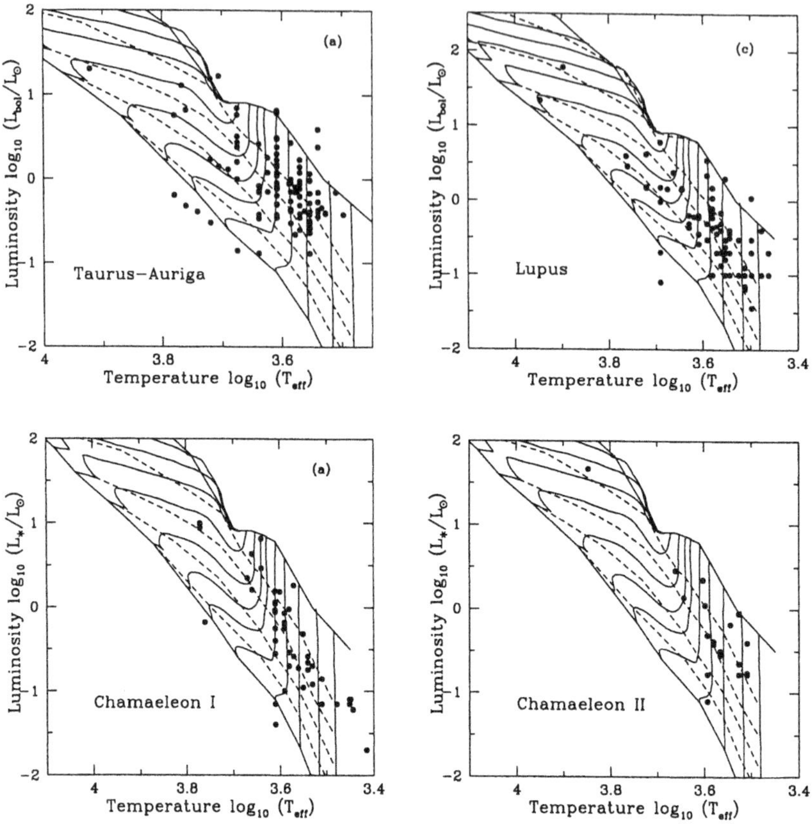

Fig. 56. The distribution of the stellar population of nearby T associations in the H-R diagram. (From Palla & Stahler [43])

Chamaeleon The Chamaeleon SFR consists of three molecular clouds (Cha I, II and III), and a number of smaller structures, including compact globules and isolated dark clouds (Schwartz [82]). Because of their location at high galactic latitude ($b \approx -17°$), the clouds suffer from little contamination by background sources. Of the three dark clouds, the Cha I region has the highest number of PMS stars, the majority of which are optically visible, with very few Class I sources. Cha II also contains a significant population of young stars, whereas no visible or infrared sources have been discovered in Cha III. The distance to the clouds was, until recently, poorly known, ranging from 115 to 215 pc. However, recent estimates establish a value of $d = 160$ pc for Cha I and $d = 180$ pc for Cha II (Whittet et al. [100]).

In Cha I, the stellar census derived from optical, near-infrared and X-ray surveys consists of about 100 probable members, with roughly 50% being WTTS (e.g. Persi et al. [75]). The stellar population of Cha II contains about 20 stars, whose properties have been determined by Hughes & Hartigan [45].

The H-R diagram of these two regions are displayed in Fig. 56. Note the presence in Cha I of a number of stars with mass in the brown dwarf regime: spectroscopic studies by Neuhäuser & Comerón [69] have confirmed the substellar nature for four of them. The ROSAT survey has discovered about 80 additional young stars spread over a wide area (Alcalá et al. [2]). Although about half of them appear to be PMS objects on the basis of the strength of the lithium line, their genetic link to the central clouds is still unclear (Terranegra et al. [91]).

ρ Ophiuchi The ρ Ophiuchi complex, at a distance of 160 pc, represents another region of intense star formation. The two main clouds, L1688 and L1689, are massive and opaque, with visual extinction exceeding 50 mags (Wilking & Lada [103]). Most of the young stars are found in the L1688 cloud core, where near-infrared surveys have revealed a population of over 200 heavily embedded objects (e.g. Barsony et al. [6]). Because of the high extinction, stellar parameters are difficult to determine. This task has been successfully done by means of near-infrared spectroscopy for a sample of about 100 stars, located in the central region of L1688 (Greene & Meyer [34]; Luhman & Rieke [56]). The distribution of 72 stars from this group in the H-R diagram is also shown in Fig. 57. The population is nicely spread below the birthline, and, once again, there is evidence for a number of brown dwarfs (e.g. Wilking et al. [104]).

Unlike the other SFRs discussed above, there is little sign of a dispersed population of older stars in ρ Oph. Of the 35 RASS stars studied by Martín et al. [60], only three are post-T Tauri stars, while there is an equal number of CTTS and WTTS within the interior of L1688.

Upper Scorpius Upper Scorpius is the youngest subgroup of the Scorpius-Centaurus OB association, the nearest to the Sun at a distance of 145 pc. From proper motion studies, Blaauw [8] derived an age of 2×10^7 yr for the whole complex, while the kinematic age of Upper Scorpius is closer to 5×10^6 yr. The whole area is essentially free of molecular gas, and the absence of deeply embedded objects indicates that there is little star formation activity going on in the association. The Upper Scorpius association has been thoroughly investigated by Hipparcos which has established secure kinematic membership for 120 stars (de Zeeuw et al. [107]). The stellar properties of the intermediate- and high-mass members have been derived by de Geus et al. [30].

The low-mass population associated with Upper Scorpius has been studied mostly by follow-up studies of X-ray observations, both with EINSTEIN (Walter et al. [99]) and ROSAT (Sciortino et al. [83]; Preibisch & Zinnecker [77]). The H-R diagram of Fig. 57 includes both high- and low-mass stars. Note how the majority of the intermediate mass stars lie close to the ZAMS, while the brightest stars depart from it. As for the PMS population, most

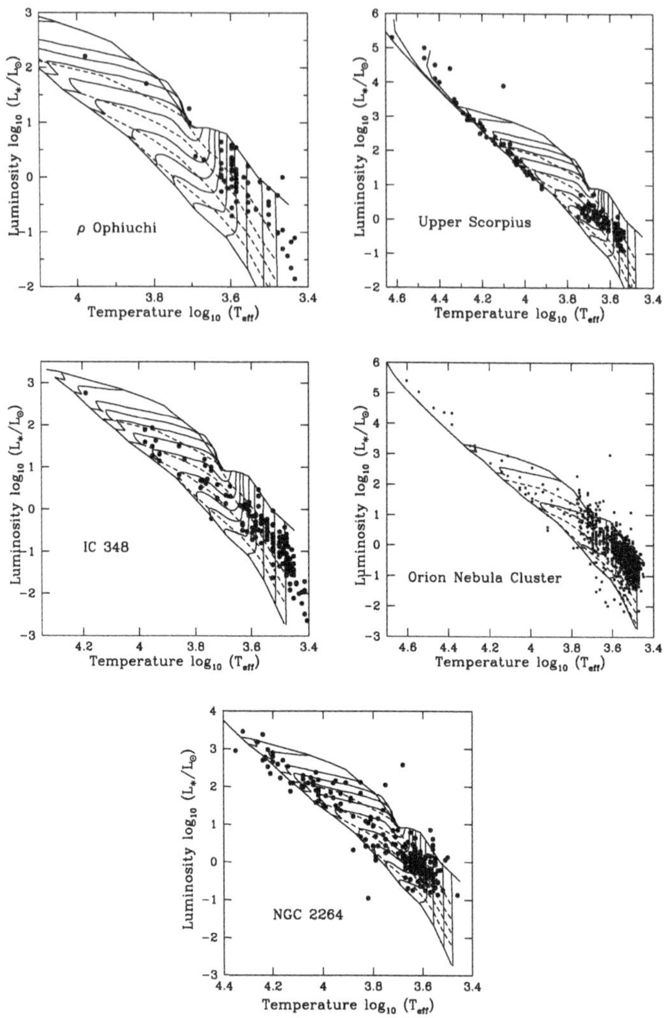

Fig. 57. The distribution of the stellar population of nearby clusters in the H-R diagram. (From Palla & Stahler [43])

of the stars are WTTS, with a small number of older post-T Tauri objects. The gap between the two distributions is not real, and results from selection effects in the definition of the samples. Clearly, the complete population of Upper Scorpius should include stars of both higher and lower luminosities and must still be fully characterized.

IC 348 IC 348 is a small cluster in the Perseus complex, a large region of high visual extinction at a distance of 320 pc. In addition to IC 348, the area contains other well known SFRs, such as L1448, L1455, B1, B5, and

NGC 1333 (Bachiller & Cernicharo [4]). The richest groups of PMS stars are found in IC 348 and NGC 1333, which contain bright B type stars as well as a cluster of deeply embedded Class 0 sources. The entire IC 348 cluster is embedded within a dark cloud of diameter of 2 pc.

The stellar population of IC 348 has been extensively studied over a range of wavelengths. In X-rays, deep ROSAT observations have identified more than 100 sources, 56 of them probable new WTTS (Preibisch et al. [78]). In the optical, the stellar census has been extended to more than 110 emission line stars (Herbig [38]), whereas deep near-infrared imaging has revealed about 400 probable members (Lada & Lada [52]). Follow-up optical and infrared spectroscopy has allowed to derive stellar parameters for more than 150 stars (Luhman et al. [57]; Luhman [55]). The distribution of these stars is displayed in Fig. 57. The brightest member of the cluster is BD+31 643, a B5 star that, together with other 11 similar objects, belongs to the expanding Per OB2 association.

Orion Nebular Cluster The Trapezium cluster was first noted by Trumpler in 1931 and Baade & Minkovski in 1937 as a concentration of faint red stars near the $^1\theta$ Ori Ic association. From optical studies, Herbig & Terndrup [39] identified about 150 stars over a region approximately $3' \times 3'$ in size around the Trapezium stars. Deep optical and near-infrared imaging has revealed the full population of ~3000 low-mass members, equally proportioned between embedded and revealed stars (McCaughrean & Stauffer [61]). In the central 1 arcmin (0.14 pc) diameter core, the stellar number density exceeds 10^4 pc^{-3}, with a mean separation between the cluster stars of ~0.045 pc. At a distance of only 470 pc, the Trapezium cluster is one of the densest stellar concentrations in the Galaxy.

The study of Hillenbrand [23] has provided stellar parameters for more than 900 visible stars. Their distribution in the H-R diagram is shown in Fig. 57. Most stars of low- and intermediate-mass fall below the birthline, while those of mass larger than ~6 $M_\odot$ have already joined the ZAMS. The preponderance of low-mass stars is clearly visible in the H-R diagram. Hillenbrand found that the actual mass distribution is roughly a power law above 0.2 $M_\odot$, and also noted a turnover at lower masses. Incompleteness at the low-mass end does not allow, however, to draw firm conclusion about the shape of the IMF. Finally, from the tight distribution of stars near the birthline, it is clear that the period of most active formation is confined to a few $\times 10^6$ yr, and has recently ended with gas dispersal from the Trapezium (although formation is going on inside the molecular cloud behind the cluster).

NGC 2264 At a distance of 750 pc, NGC 2264 is the central cluster of the Mon OB1 association, which contains a large number of reflection nebulae and dark clouds. This cluster represents a benchmark for the observational studies of star formation, starting with the initial discovery of 84 Hα emission line stars by Herbig [36]. Subsequent photometric studies by Walker [97] showed the stars to be PMS in nature. Walker concluded that all stars with spectral type later than A0 lie above the ZAMS, resulting in an age estimate for the cluster of $\sim 5 \times 10^6$ yr. Several hundred optically visible members are now known, ranging in mass from the O7 star S Mon to T Tauri stars.

The small extinction and differential reddening, and the presence of an optically thick reflection nebula which obscures background stars greatly facilitate the identification of cluster members (Pérez et al. [74]). However, until recently the number of stars with reliable values of luminosity and effective temperature has remained limited. Now, optical and X-ray surveys have been presented by Sung et al. [90] and Flaccomio et al. [25] [26], determining the properties of about 200 low-mass stars. The distribution of these stars and of main-sequence stars with masses greater than 1.5 $M_\odot$ in the H-R diagram is shown in Fig. 57. In spite of the large distance, we see that the low-mass population has been sampled down to $\sim$0.2 $M_\odot$.

13.2 Star Formation Histories

Having established the stellar content of the most conspicuous star forming units within $\sim$1 kpc from the Sun, we can now utilize the evolutionary tracks and isochrones to gauge the formation histories by reading off individual ages of their members. This exercise produces age histograms which reveal the pattern of stellar births as a function of time. The results for associations and clusters are shown in Fig. 58 and Fig. 59. Note that only stars with ages less than 10^7 yr are included in the figures, even though the H-R diagrams presented above indicate that older stars are also present in each region. However, their number is so small and the age dispersion is so large that their inclusion does not add much information, nor does it change the shape of the distribution.

Scanning the histograms from the oldest ages to the youngest, we see a common character of star formation: it begins at a modest level roughly 4×10^6 yr in the past and is followed by a rapid rise toward the present epoch. This qualitative feature is insensitive to the precise method of assessing stellar luminosities and effective temperatures. In the Orion Cluster, however, the acceleration of star formation is a factor of at least 10 higher than in the other cases. Only in the case of Upper Scorpius, we see that at about 3×10^6 yr the formation rate peaked and then underwent a steep decline. This falloff is also evident in the H-R diagram, as the gap just below the birthline and the clumps of PMS stars. The lack of molecular gas is clearly responsible for the cessation of active star formation in Upper Scorpius, an event possibly related to the explosion of a supernova (see de Geus [29], Preibisch & Zinnecker [77]).

In order to quantify the observed trend, Table 5 lists the e-folding times $t_\circ$ of the accelerating formation. These times fall between 1 and 3×10^6 yr, an indication of the brevity of the process. While for regions containing massive

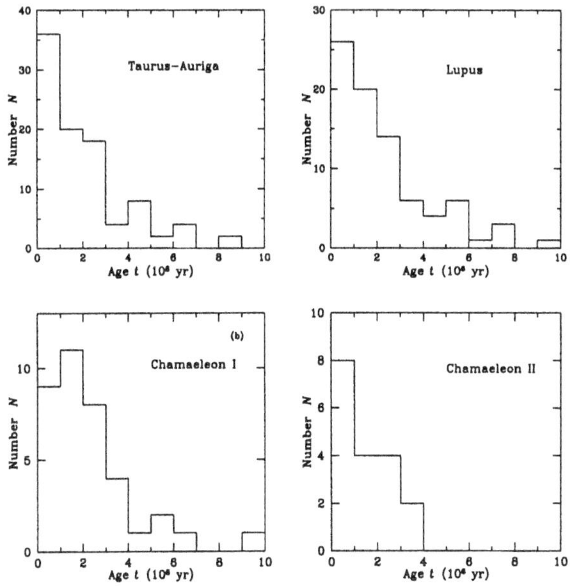

Fig. 58. The distribution of the stellar population of nearby T associations in the H-R diagram. (From Palla & Stahler [43])

Table 5. Formation Time Scales

Region	Distance (pc)	Time t_o (10^6 yr)
Taurus-Auriga	140	2.3
Lupus	140	2.3
Chamaeleon I	160	1.5
Chamaeleon II	180	1.8
ρ Ophiuchi	160	1.0
Upper Scorpius	150	1.8
IC 348	320	1.0
Orion Nebula	470	1.2
NGC 2264	760	3.3

stars, the activity should soon end because of the efficient dispersal of the remaining molecular gas by radiative and mechanical forces, in T associations where such type of stars are absent, the acceleration can be maintained for another few million years, but no longer. The scarcity of stars with ages of

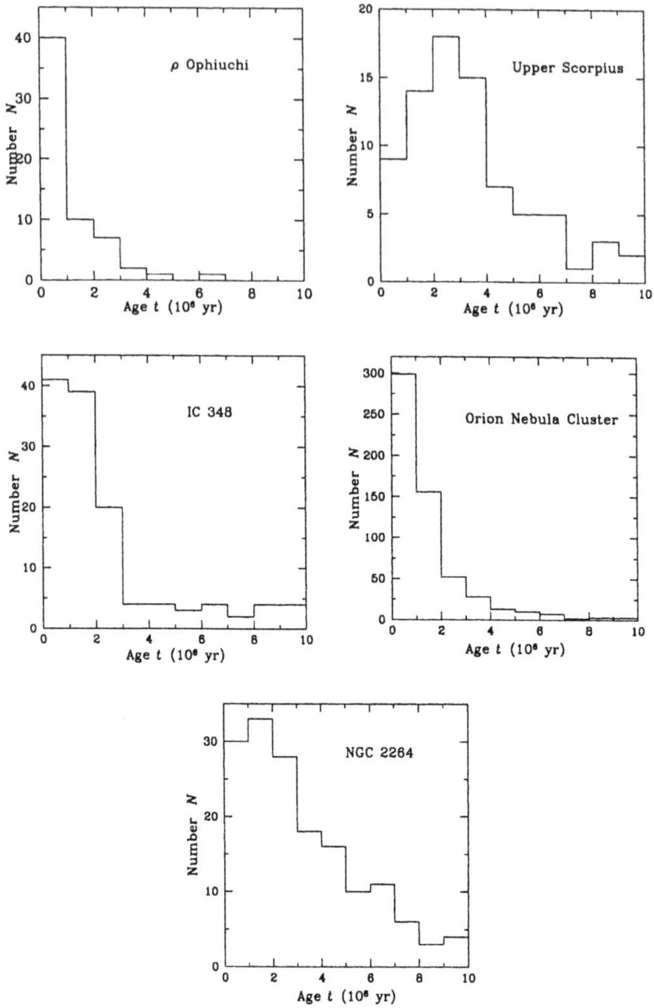

Fig. 59. The distribution of the stellar population of nearby young clusters in the H-R diagram. (From Palla & Stahler [43])

10^7 yr or older in fact implies that those regions cannot remain for that length of time in a state of active formation, and eventually star formation must stop. However, the physical mechanism(s) responsible for the decline is not obvious, since winds and outflows from low-mass stars are too inefficient to disperse the gas in just a few million years.

How robust is the accelerating pattern? Does it depend on the particular choice of PMS tracks and isochrones used for assessing stellar ages? Would factors such as residual mass accretion, the location of the birthline, or the presence of unresolved binaries affect the resulting age distribution and wipe out the non-steady nature of formation within each cluster/association? Clearly, all these effects should be present on an individual basis and must be accounted for when more complete observations become available. However, we believe that accelerating star formation will remain as observations improve (for a discussion on the assessment of uncertainties, see Palla & Stahler [43]). In support of this conclusion, it is interesting to consider the extreme view in which star formation indeed occurs steadily for 10^7 yr: how would the H-R diagram of such a hypothetical region look like? Using the data base of Taurus-Auriga, Fig. 60 shows the resulting H-R diagram obtained by assigning random ages between 0 and 10^7 yr to the stars distributed between the birthline and the ZAMS. Compared to the original H-R diagram, the faked one is markedly different with a large void just below the birthline and a concentration of stars around the last few isochrones. Such a diagram resembles that of Upper Scorpius where the absence of molecular gas indicates that star formation has halted. Thus, we reach the conclusion that regions with still considerable amount of molecular gas always have enough young objects to produce an accelerating rate.

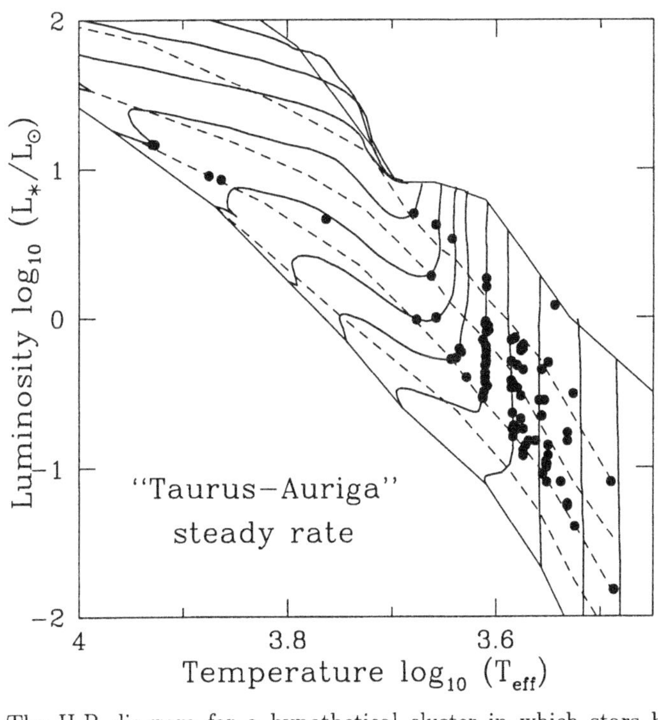

Fig. 60. The H-R diagram for a hypothetical cluster in which stars have been randomly assigned ages between 0 and 10^7 yr. (From Palla & Stahler [43])

13.3 The Post-T Tauri Problem

The rapid rise of star formation sheds new light on the post-T Tauri problem. Herbig [31] first noted that there is a remarkable deficit of T Tauri stars of age greater than approximately 3 million years in nearby star forming regions. He introduced the possibility of the existence of a missing population of post-T Tauri stars with ages greater than $\sim 5 - 10 \times 10^6$ yr. Searches aimed at identifying such an elusive population have been carried out ever since. However, the issue is still matter of heated debate. I believe that the implications bear important consequences for the understanding of the star formation process and therefore deserve some further remarks.

Observational Considerations The main question is: does the combination of CTTS and WTTS observed *within* or *close* to molecular cloud complexes represent the whole stellar population spawned by such entities? If so, the history of star formation can be traced accurately by their stellar content. This approach has been followed in the previous sections (see also the work of Briceño and collaborators [17] [11]). In contrast, other authors have remarked that the dominant stellar population has still to be determined, since many relatively young stars can be found at large distances from the parent molecular clouds. Since the measured velocity dispersion of young stars is $\sim$2-5 km s^{-1}, one expects that they can drift over 5–20 pc in $\sim 10^7$ yr (equivalent to $\sim 8 - 20°$ at 150 pc), ignoring the gas content and assuming ballistic trajectories.

The finding of hundreds of RASS stars over large areas of sky (much larger than those probed by optical/near-infrared surveys) has boosted this idea and rejuvenated the debate on the post-T Tauri stars. If these stars are genetically linked to those found within the boundaries of molecular clouds, then the total number of low-mass stars formed in such regions would increase dramatically, filling the gap shown by the age distributions shown in Figs. 58 and 59. In particular, Feigelson [24] estimated that the number of stars older than $2 - 5 \times 10^6$ yr should exceed that of younger stars by factor 5–10 or larger, assuming a constant rate of star formation over the last $\sim 10^7$ yr. As a consequence, the star formation efficiency of molecular clouds would be much higher than the few percent currently estimated.

The nature and origin of the widely dispersed population of T Tauri stars are still uncertain. To complicate things even more is the unknown distance of the majority of these stars. The presence of strong lithium absorption line observed at high spectral resolution in a number of the RASS stars clearly indicates that they are young objects (e.g. Covino et al. [18]) and the additional information on the distance provided by Hipparcos has allowed to place some of them in the H-R diagram, confirming the PMS nature (Neuhäuser & Brandner [68]). However, the number of stars with accurate distance measurements is too small to draw general conclusions. Also, the use of the strength of the lithium line to distinguish PMS from main sequence stars is not free of uncertainty and, in fact, Briceño et al. [10] have argued that most of the RASS stars of spectral type G and early K (to which the RASS is most sensitive) must have ages closer to 10^8 yr than 10^7 yr. Quantitative analysis of ages from lithium depletion also provides little evidence for a substantial population of post-T Tauri stars (Favata et al. [23] and Martín & Magazzù [59]).

In conclusion, it appears that while there is a number of young PMS stars at large distances from star forming complexes, the majority of the RASS represent an older population, likely formed in clouds of a previous generation which have been dispersed by now. Various possibilities have been suggested to explain their origin (see Neuhäuser [67]). The most interesting one links the RASS stars to the Gould Belt, a band of O, B and early A type stars highly inclined to the galactic plane. A number of star forming regions are aligned within this band (Lupus, Orion, Sco-Cen), while others are well outside (Taurus, Chamaeleon). The RASS stars found in these latter regions could have formed in low-mass cloudlets (Feigelson [24]) or could represent run-away stars, ejected from the birthsites as a result of dynamical interactions (Sterzik et al. [89]).

Theoretical Considerations In order to explain the post-T Tauri problem, two possible solutions have been suggested. In both models, the lack of older stars is a consequence of the star formation process itself rather than the dispersal/ejection mechanisms mentioned above. However, in one case star formation occurs in an accelerating pattern after a long period of quiescence (Palla & Galli [73]; Palla & Stahler [43]), while in the other case star formation is a very rapid, dynamic process which only lasts for a few million years (Ballesteros-Paredes et al. [5]; Elmegreen [20]).

Star Formation Is Slow... The basic idea here is that it is incorrect to think of star formation as an instantaneous process during the lifetime of a molecular cloud. Rather, it takes a long time to establish the appropriate initial conditions for gravitational collapse and protostellar accretion. In the magnetically supported clouds discussed in Sect. 2, the time scale is set by the ambipolar diffusion time (see eq. (19), which is similar to the estimated lifetime of molecular clouds ($\gtrsim 10^7$ yr). Dense clumps within larger clouds contract quasi-statically before the density exceeds some critical threshold. Both the accretion and the core dispersal time scales are much shorter than the cloud lifetime. Therefore,

- the probability of finding a large population of stars with ages greater than $\sim 10^7$ yr is *intrinsically* very small;
- the star formation rate is *not* constant as a function of time, but very low initially and steeply *accelerating* at later times;
- the overall star formation efficiency is small (few percent) because of the fast *deceleration* related to the mechanisms that determine the survival

of a molecular clump (i.e. negative *feedback* due to outflows and injection of energy from the newly formed stars);
- star formation needs not to be synchronized to small age spreads ($\sim 1-2\times 10^6$ yr), since it takes place at all times during the life of a molecular cloud, albeit at different rates;
- giant molecular complexes contain so many of these clumps that at least few of them have evolved to the point of initiating star formation, thus explaining why almost all nearby molecular clouds are forming stars.

...or Star Formation Is Fast In this picture, star formation is not mediated by any retarding force, but occurs in only one or two crossing times for a wide range of scales, from $\sim$1 to $\sim 10^3$ pc. Observational evidence in support of this claim has been summarized by Elmegreen [20]. This conclusion runs opposite to the long believed longevity of molecular clouds, based on the disparity between the free fall time and the total star formation rate in Galactic clouds. Recent MHD simulations of self-gravitating clouds find that turbulence decays rapidly over a dynamical time (e.g. MacLow [58]), eliminating the need for sustaining the clouds for many crossing times. However, one must explain how to create continuously new molecular clouds at the conditions appropriate for fast collapse. Since self-gravity of the gas cannot operate so rapidly, one must invoke external factors as triggering events. In the case of Taurus-Auriga, Ballesteros-Paredes et al. [5] have simulated the formation of clouds resulting from the compression of HI flows. Some general implications of the picture include:

- the formation of stars occurs only in self-gravitating cores, and the critical factor is not their high densities;
- because of the hierarchical structure of molecular clouds, such conditions are rarely found, thus explaining the low overall efficiency, both locally and over a Galactic scale;
- because of the short time scale, young stars do not have time to move away form their birthsites, maintaining the turbulent-driven structure of the parent clouds.

Conclusion

Differences in the two pictures of star formation (*slow vs. rapid*) are so large that observational tests should be able to discriminate between them. Future studies should try to specifically address the predictions of these models. In the meantime, we can conclude with the following consideration. We have started this cycle of lectures underlining how the initial conditions established during protostellar accretion have modified our view of PMS evolution with respect to the classical studies of Hayashi, Iben, Cameron and their collaborators. We have now reached the point where the global properties of PMS stars in nearby clusters and associations are providing fundamental tests to the theory of star formation. The circle is closed.

References

1. Adams, F.C., Fatuzzo, M. 1996, ApJ, 464, 256
2. Alcalá, J.M., Krautter, J., Schmittt, J.H.M.M., Covino, E., Wichmann, R., Mundt, R. 1995, A&AS, 114, 109
3. André P., Ward-Thompson D., Barsony M. 1993, ApJ, 406, 122
4. Bachiller, R., Cernicharo, J. 1986, A&A, 166, 283
5. Ballesteros-Paredes, J., Hartmann, L., Vázquez-Semadeni, E. 1999, ApJ, 527, 285
6. Barsony, M., Kenyon, S.J., Lada, E.A., Teuben, P.J. 1997, ApJS, 112, 109
7. Bate, M.R., Clarke, C.J., McCaughrean, M.J. 1998, MNRAS, 297, 1163
8. Blaauw, A. 1964, ARAA, 2, 213
9. Bonnell I., Davies M.B. 1998, MNRAS, 295, 691
10. Briceño, C., Hartmann, L.W., Stauffer, J.R., Gagné, M., Stern, R.A., Caillault, J.P. 1997, AJ, 113, 740
11. Briceño, C., Calvet, N., Kenyon, S., Hartmann, L. 1999, AJ, 118, 1354
12. Carpenter, J.M., Meyer, M.R., Dougados, C., Strom, S.E., Hillenbrand, L.A. 1997, AJ, 114, 198
13. Cesaroni R., Churchwell E., Hofner P., Walmsley C.M., Kurtz S. 1994, A&A, 288, 903
14. Cesaroni R., Felli M., Testi L., Olmi L., Walmsley C.M. 1997, A&A, 325, 725
15. Cesaroni, R., Hofner, P., Walmsley, C.M., Churchwell, E. 1998, A&A, 331, 709
16. Clarke C., Bonnell I., Hillenbrand L.A. 2000, in Protostars and Planets IV, ed. V. Mannings, A.P. Boss & S.S. Russell (Tucson: University of Arizona Press), p. 151
17. Codella C., Testi L., Cesaroni R. 1997, A&A, 325, 282
18. Covino, E., Alcalá, J.M., Allais, S., Bouvier, J., Terranegra, L., Krautter, J. 1997, A&A, 328, 187
19. Elmegreen B.G. 1999, ApJ, 515, 323
20. Elmegreen B.G. 2000, ApJ, 530, 277
21. Elmegreen, B.G., Clemens, C. 1986, ApJ, 294, 523
22. Elmegreen B.G., Falgarone, E. 1996, ApJ, 471, 816
23. Favata, F., Micela, G., Sciortino, S. 1997, A&A, 326, 647
24. Feigelson, E.D. 1996, ApJ, 468, 306
25. Flaccomio, E., Micela, G., Sciortino, S., Favata, F., Corbally, C., Tomaney, A. 1999, A&A, 345, 521
26. Flaccomio, E., Micela, G., Sciortino, S., Damiani, F., Favata, F., Harnden, Jr.,F.R., Schachter, J. 2000, A&A, 355, 651
27. Fuente A., Martin-Pintado J., Bachiller R., Neri R., Palla F. 1998, A&A, 334, 253
28. Garay G., Lizano S. 1999, PASP, 111, 1049
29. de Geus, E.J. 1992 A&A, 262, 258
30. de Geus, E.J., de Zeeuw, P.T., Lub, J. 1989, A&A, 216, 44
31. Ghez, A.M., Neugebauer, G., Matthews, K. 1993, AJ, 106, 2005

32. Gladwin, P.P., Kitsionas, S., Boffin, H.M.J., Whitworth, A.P. 1999, MNRAS, 302, 305
33. Gomez, M., Hartmann, L., Kenyon, S.J., Hewett, R. 1993, AJ, 105, 1927
34. Greene, T.P., Meyer, M.R. 1995, ApJ, 450, 233
35. Habing H.J., Israel F.P. 1979, ARAA, 17, 345
36. Herbig, G.H. 1954, ApJ, 119, 483
37. Herbig, G.H. 1962, ApJ, 135, 736
38. Herbig, G.H. 1998, ApJ, 497, 736
39. Herbig, G.H., Terndrup, D.M. 1986, ApJ, 307, 609
40. Herbst W., Miller, D.P. 1982, AJ, 87, 1478
41. Hillenbrand L.A. 1995, Ph.D. Thesis, Univ. of Massachusetts
42. Hillenbrand L.A., Hartmann, L. 1998, ApJ, 492, 540
43. Ho, P.T.P., Haschick, A. 1981, ApJ, 309, 514
44. Hollenbach, D.J., Johnstone, D., Lizano, S., Shu, F.H. 1994, ApJ, 428, 654
45. Hughes, J., Hartigan, P. 1992, AJ, 104, 680
46. Hughes, J., Hartigan, P., Krautter, J., Kelemen, J. 1994, AJ, 108, 1071
47. Iben, I., Talbot, R.J. 1966, ApJ, 144, 968
48. Kenyon, S.J., Brown, D.I., Tout, C.A., Berlind, P. 1998, AJ, 115, 2491
49. Krautter, J., Wichmann, R., Schmitt, J.H.M.M., Alcalá, J.M., Neuhäsuer, R., Terranegra, L. 1997, A&AS, 123, 103
50. Kroupa, P. 2000, in Massive Stellar Clusters, eds. A. Lancon & C. Boily (San Francisco: ASP), p. 319
51. Kurtz S., Cesaroni R., Churchwell E., Hofner P., Walmsley C.M. 2000, in Protostars and Planets IV, ed. V. Mannings, A.P. Boss & S.S. Russell (Tucson: University of Arizona Press), p. 299
52. Lada, E.A., Lada, C.J. 1995, AJ, 109, 1682
53. Lada, C.J., Margulis, M., Dearborn, D. 1984, ApJ, 285, 141
54. Larson, R.B. 1995, MNRAS, 272, 213
55. Luhman, K.L. 1999, ApJ, 525, 440
56. Luhman, K.L., Rieke, G.H. 1999, ApJ, 525, 440
57. Luhman, K.L., Rieke, G.H., Lada, C.J., Lada, E.A. 1998, ApJ, 508, 347
58. MacLow, M.M. 1999, ApJ, 524, 169
59. Martín, E.L., Magazzù, A. 1998, A&A, 342, 173
60. Martín, E.L., Montmerle, T., Gregorio-Hetem, J., Casanova, S. 1998, MNRAS, 300, 733
61. McCaughrean, M.J., Stauffer, J.R. 1994, AJ, 108, 1382
62. Molinari S., Testi, L., Brand J., Cesaroni R., Palla F. 1998, ApJ, 505, L39
63. Motte, F., André, P, Neri, R. 1998, A&A, 336, 150
64. Nakajima, Y., Tachihara, K., Hanawa, T., Nakano, M. 1998, ApJ, 497, 721
65. Nakano, T., Hasegawa, T., Norman, C. 1995, ApJ, 450, 183
66. Natta A., Grinin V., Mannings V. 2000, in Protostars and Planets IV, ed. V. Mannings, A.P. Boss & S.S. Russell (Tucson: University of Arizona Press), p. 599
67. Neuhäuser, R. 2000, Rev. Mod. Astron., in press
68. Neuhäuser, R., Brandner, W. 1998, A&A, 330, L29
69. Neuhäuser, R., Comerón, F. 1999, A&A, 350, 612
70. Norman, C., Silk, J. 1980, ApJ, 238, 158
71. Norris, R.P., McCutcheon, W.H., Caswell, J.L., Wellington, K.J., Reynolds, J.E., Peng, R.S., Kesteven, M.J. 1998, Nature, 335, 149

72. Olmi L., Cesaroni R., Walmsley C.M. 1993, A&A, 276, 489
73. Palla, F., Galli, D. 1997, ApJ, 476, L35
74. Pérez, M.R., Thé, P.S., Westerlund, B.E. 1987, PASP, 99, 1050
75. Persi, P., Marenzi, A.R., Olofsson, G., et al. 2000, A&A, 357, 219
76. Plambeck, R.L., Wright, M.C.H., Carlstrom, J.E. 1990, ApJ, 348, L65
77. Preibisch, Th., Zinnecker, H. 1999, AJ, 117, 2381
78. Preibisch, Th., Zinnecker, H., Herbig, G.H. 1996, 310, 456
79. Raboud D., Mermilliod J.-C. 1998, A&A, 333, 897
80. Reid, M.J., Haschick, A., Burke, B.F., Moran, J.M., Johnston, K.J., Swenson, G.W. 1981, ApJ, 239, 89
81. Schwartz, R.D. 1977, ApJS, 35, 161
82. Schwartz, R.D. 1991, in Low Mass Star Formation in Southern Molecular Clouds, ESO Report No. 11, ed. B. Reipurth, p. 93
83. Sciortino, S., Damiani, F., Favata, F., Micela, G. 1998, A&A, 332, 825
84. Shepherd D.S., Churchwell E. 1996, ApJ, 472, 225
85. Simon, M. 1997, ApJ, 482, L81
86. Stahler, S.W. 1985, ApJ,ApJ, 293, 207
87. Stahler S.W., Palla F., Ho P.T.P. 2000, in Protostars and Planets IV, ed. V. Mannings, A.P. Boss & S.S. Russell (Tucson: University of Arizona Press), p. 327
88. Stecklum, B., et al. 1998, First Circumstellar Disk around a Massive Star (ESO Press Release PR-08-98)
89. Sterzik, M., Alcalá, J.M., Neuhäuser, R., Schmitt, J.H.M.M. 1995, A&A, 297, 418
90. Sung, H. Bessell, M.S., Lee, S.W. 1997, AJ, 114, 2644
91. Terranegra, L., Morale, F., Spagna, A., Massone, G., Lattanzi, M.G. 1999, A&A, 341, L79
92. Testi, L., Sargent, A.I. 1998, ApJ, 508, L91
93. Testi L., Palla F., Natta A. 1999, A&A, 342, 515
94. Testi L., Palla F., Prusti, T., Natta, A., Maltagliati, S. 1997, A&A, 320, 159
95. Ungerechts, H., Thaddeus, P. 1987, ApJS, 63, 645
96. van Buren, D., Mac Low, M.-M., Wood, D.O.S., Churchwell, E. 1990, ApJ, 353, 570
97. Walker, M.F. 1956, ApJS, 2, 365
98. Walter, F.M., Brown, A., Mathieu, R.D., Myers, P.C., Vrba, F.V. 1988, AJ, 96, 297
99. Walter, F.M., Vrba, F.J., Mathieu, R.D., Brown, A., Myers, P.C. 1994, AJ, 107, 692
100. Whittet, D.C.B., Prusti, T., Franco, G.A.P., Gerakines, P.A., Kilkenny, D., Larson, K.A., Wesselius, P.R. 1997, A&A, 327, 1194
101. Wichmann, R., Krautter, J., Schmitt, J.H.M.M. et al. 1996, A&A, 312, 439
102. Wichmann, R., Covino, E., Alcalá, J.M., Krautter, J., Allain, S., Hauschildt, P.H. 1999, MNRAS, 307, 909
103. Wilking, B.A., Lada, C.J. 1983, ApJ, 274, 698
104. Wilking, B.A., Greene, T.P., Meyer, M.R. 1997, AJ, 117, 469
105. Wood, D.O.S., Churchwell, E. 1989, ApJ, 340, 265
106. Yorke, H. 1984 in Workshop on Star Formation, ed. R.D. Wolstencroft (Edinburgh: Royal Obs.), 63
107. de Zeeuw, P.T., Hoogerwerf, R., de Bruijne, J.H.J., Brown, A.G.A., Blaauw, A. 1999, AJ, 117, 354
108. Zhang, Q., Hunter, T.R., Sridharan, T.K. 1998, ApJ, 505, L151

Observations of Young Stellar Objects (YSO)

Hans Zinnecker

Astrophysikalisches Institut Potsdam
An der Sternwarte 16
14482 Potsdam, Germany

I. Introduction

This lecture course is unusual in that it consists of a body of real observing proposals, submitted to various time allocation committees in the years 1996–1999. In all cases the observations have already been conducted; a summary and reference to published results are given at the end of each proposal. I am grateful to my co-investigators B. Brandl, W. Brandner, A. Brown, M. J. McCaughrean, Th. Preibisch, B. Reipurth, Th. Stanke, K. Stapelfeldt, and A. Quirrenbach for their permission to use these proposals.

The organization of this chapter is as follows: The first part (5 lectures) mainly deals with disks, jets, and protostars. In particular, the first two lectures are on disks and jets and the next two are on the search for jet sources and protostars. Lecture 5 is on the circumstellar environment of weak-line or seemingly naked T Tauri stars, relevant to planet formation. The second part (3 lectures) starts with a discussion of the statistics of young binary stars, a subject that is especially dear to my heart (*cf.* the recent IAU Symp. 200). It continues with a study of the binarity of young A and B stars and ends with a study of the multiplicity of massive stars. The third part (4 lectures) basically deals with the low-mass stellar content of OB associations and young clusters. We begin with a discussion of how to determine the low-mass IMF in the most nearby OB association (from ground-based observations) and end with our quest for the low-mass pre-Main Sequence population in the famous 30 Dor starburst cluster in the LMC (from HST observations). In between we discuss the stellar content of two important young galactic clusters, one small (IC 348) and one big (NGC 3603).

This course is dedicated to George Herbig, on the occasion of his 80th birthday. George very kindly agreed to provide a historical introduction to this Saas-Fee course about the early history of star formation for which I would like to thank him very much.

I also thank U. Hanschur for his invaluable technical help in preparing my chapter at long last. Finally, I thank the organisers and editors, A. Maeder and G. Meynet, for inviting me to give this course and for their patience with the manuscript.

1 How to Find Young Stars and Protostars

The poster for this Saas-Fee course is entitled "Physics of Star Formation in Galaxies" and shows an optical image of the NGC 604 HII region in the Triangulum galaxy M33 as an example. NGC 604 is one of the most luminous HII regions in the Local Group, second only to 30 Doradus in the Large Magellanic Cloud [18].

However, much of this course will be concerned with star formation on smaller scales, scales that can only be studied nearer to home, *i.e.* inside our own Milky Way Galaxy or even only close to the solar neighborhood (< 500 pc distance), due to sensitivity and spatial resolution. The most well-known nearby star forming regions include the Taurus-Auriga, Rho Ophiuchi, Lupus, R Coronae Australis, and Chamaeleon T associations which are all at a distance of about 150 pc (all but Taurus-Auriga are located in the southern sky, see [17]). Moreover, there are nearby OB associations [3], the nearest being the Scorpius-Centaurus (Sco-Cen) OB association, also at about 150 pc, and the most famous being the Orion OB association at 350 – 500 pc (see [4]). Orion is distinct in that it is the place in the sky where massive stars can be studied in the making. On the other hand, in the Sco-Cen region star formation has just finished, and thus this young association is a very good place where the end-products of the star formation process - *e.g.* the stellar mass distribution - can be investigated.

Detailled observations of true protostars in the process of collapse (so-called class 0 objects) have also become possible recently, owing to technological advances in radio astronomy, such as new mm receivers and mm optimised telescopes, including telescope arrays for mm interferometry (at IRAM, OVRO, BIMA). Before, beam dilution prevented sensitive, spatially resolved mm spectroscopy of infalling gas. Now, the hallmark of a collapsing protostar is an asymmetric line profile in an optically thick tracer molecule of dense gas (such as CS in its various rotational transitions), with a red-shifted self-absorption feature [see Fig. 1]; such profiles look very much like the shape of the Matterhorn mountain (as we saw it during our excursion).

In the following, to conclude this brief Introduction, a quick tabular overview is given of the various techniques available to find young stars near their birthplaces (molecular clouds), followed by a few references to young star surveys.

techniques:	$\Longrightarrow$	**objectives:**
CO surveys (2.6 mm, 1.3 mm)	$\Longrightarrow$	molecular clouds
CS surveys (3.0 mm, 2.0 mm)	$\Longrightarrow$	dense gas cores
mm continuum surveys	$\Longrightarrow$	cold dust condensations
IRAS survey (12 / 25 / 60 / 100 μm)	$\Longrightarrow$	dusty protostars
NIR surveys	$\Longrightarrow$	young objects
DENIS (IJK) / 2MASS (JHK)		with infrared excess

H_2 (2.12μm) surveys	$\Longrightarrow$	very young jets
optical surveys (Hα, LiI, CaII)	$\Longrightarrow$	T Tauri stars
X-ray surveys (ROSAT, XMM, AXAF)	$\Longrightarrow$	T Tauri stars embedded stars
proper motion surveys (Hipparcos, ...)	$\Longrightarrow$	cluster / association membership
radio cont. surveys	$\Longrightarrow$	HII regions
maser surveys (H_2O, OH, CH_3OH)	$\Longrightarrow$	high-mass SF regions (massive protostars)
VLA/VLBA surveys (3.6 cm, 6 cm)	$\Longrightarrow$	non-thermal young sources (reconnection, magn. field)
shock-excited nebulosity	$\Longrightarrow$	Herbig-Haro objects
reflection nebulosity	$\Longrightarrow$	Herbig Ae/Be stars
emission nebulosity Hα, Hβ, [OIII]	$\Longrightarrow$	HII region, young clusters

Examples of extensive surveys for dense molecular gas include the observations of Tatematsu et al. [22], [23] in the Orion A cloud and of Lada et al. [12] in the Orion B cloud (see also Bally et al. [1]). A very important IRAS survey was Beichman et al. [2], while the first 1.3mm pre-stellar dust continuum surveys were Mezger et al. [14] and recently Motte et al. [15], Testi et al. [24], and Chini et al. [5]. Near-infrared surveys for young embedded clusters were reviewed by Zinnecker et al. [30], and the first H_2 survey for embedded protostellar jets was done by Stanke [21]. A comprehensive electronic catalog of Herbig-Haro objects is given by Reipurth (http://casa.colorado.edu/hhcat), also the author of the so-called Star Formation Newsletter.

A big survey for classical T Tauri stars was the Kiso Schmidt Hα survey (Kogure et al. [10]). The ROSAT all-sky X-ray survey in search for the widespread population of weak-line T Tauri stars was reviewed by Neuhäuser [16]; see also Guillout et al. [9] for the results of an X-ray survey along Gould's Belt. The classical Hα survey for HII regions is Sharpless [19], but a more recent work on compact/ultracompact HII regions from thermal radio continuum observations is Wood and Churchwell [28]; see also Kurtz et al. [11]. Non-thermal radio searches for young stars are discussed in Montmerle and Feigelson [7]. A survey of methanol maser emission from (high-mass) star forming regions can be found in Walsh et al. [25], while OH and H_2O masers are reviewed in Garay and Lizano [8]. In a few cases H_2O masers are also associated with low-mass protostars (*e.g.* IRAS 05413-0104 / HH 212, see Claussen et al. [6]). Garay and Lizano [8] as well as Stahler et al. [20] give an excellent summary of our current knowledge of the physical conditions of massive star formation. The nearest massive protostar is IRc2-I in the

Orion BN/KL nebula, a compact radio continuum source, which is the only luminous young object known to be associated with SiO maser emission [13].

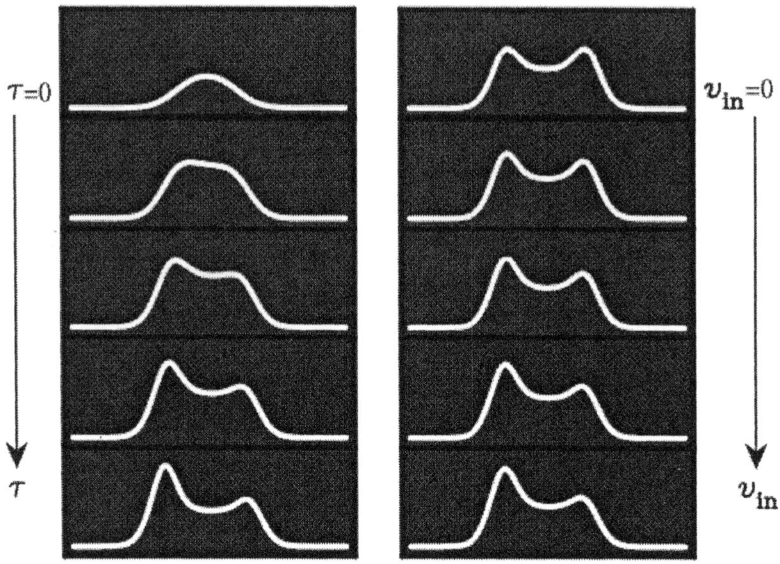

Fig. 1. Predicted molecular line profiles (intensity vs. velocity) from a spherically symmetric collapsing cloud core, as a function of the line optical depth (left) and (right) for a given optically thick line as a function of the radial infall velocity (assumed const.). Note the characteristic red-shifted self-absorption. Courtesy of Jonathan Williams (Univ. of Florida), see Williams and Myers 1999 [27], see also Wang et al. 1995 [26] and Zhou et al. 1993 [29] (especially their Fig. 8).

References

1. J. Bally, A. A. Stark, R. W. Wilson, W. D. Langer: ApJ **312**, L45 (1987)
2. C. A. Beichman, P. C. Myers, J. P. Emerson et al.: ApJ **307**, 337 (1986)
3. A. Blaauw: 'OB associations and the fossil record of star formation'. In: *The physics of star formation and early stallar evolution* ed. by Ch. J. Lada, N. D. Kylafis (Kluwer Academic Publishers, 1991) p. 125
4. A. G. A. Brown, A. Blaauw, R. Hoogerwerf, J. H. J. deBruijne, P. T. deZeeuw: 'OB associations'. In: *The Origin of Stars and Planetary Systems* ed. by Ch. J. Lada, N. D. Kylafis (Kluwer Academic Publishers, 1999) p. 411
5. R. Chini, B. Reipurth, D. Ward-Thompson et al.: ApJ **474**, L135 (1997)
6. M. J. Claussen, K. B. Marvel, A. Wootten, B. A. Wilking: ApJ **507**, L79 (1998)
7. E. D. Feigelson, Th. Montmerle: ARA&A **37**, 363 (1999)

8. G. Garay, S. Lizano: PASP **111**, 1049 (1999)
9. Guillout, J. H. M. M. Schmitt, D. Egret et al.: A&A **351**, 1003 (1999)
10. T. Kogure, S. Yoshida, S. D. Wiramihardja et al.: PASJ **41**, 1195 (1989)
11. S. Kurtz, E. Churchwell, D. O. S. Wood: ApJS **91**, 659 (1994)
12. E. A. Lada, J. Bally, A. A. Stark: ApJ **368**, 432 (1991)
13. K. M. Menten, M. J. Reid: ApJ **445**, 157 (1995)
14. P. G. Mezger, R. Zylka, J. E. Wink: A&A **228**, 95 (1990)
15. F. Motte, P. André, R. Neri: A&A **336**, 150 (1998)
16. R. Neuhäuser: Science **276**, 1363 (1997)
17. B. Reipurth: *Low mass star formation in southern molecular clouds* ed. by B. Reipurth (ESO: ESO Garching 1991)
18. M. Rosa, S. Dodorico: 'The exciting stars of giant extragalactic H II regions'. In: *Luminous stars and associations in galaxies; Proceedings of the Symposium, Porto-Kheli, Greece, May 26-31, 1985* (D. Reidel Publishing Co., Dordrecht, 1986) p. 355
19. S. Sharpless: ApJS **4**, 257 (1959)
20. S. W. Stahler, F. Palla, P. T. P. Ho: 'The Formation of Massive Stars'. In: *Protostars and Planets IV*, ed. by V. Mannings, A. P. Boss, S. S. Russell, (University of Arizona Press, Tucson, 2000) p. 327
21. Th. Stanke: An unbiased infrared H2 search for embedded flows from young stars in Orion A. PhD Thesis, Universität Potsdam (2000)
22. K. Tatematsu, T. Umemoto, O. Kameya et al.: ApJ **404**, 643 (1993)
23. K. Tatematsu, T. Umemoto, M. H. Heyer et al.: ApJS **118**, 517 (1998)
24. L. Testi, A. I. Sargent: ApJ **508**, L91 (1998)
25. A. J. Walsh, M. G. Burton, A. R. Hyland, G. Robinson: MNRAS **301**, 640 (1998)
26. Y. Wang, N. J. Evans, Sh. Zhou, D. P. Clemens: ApJ **454**, 217 (1995)
27. J. P. Williams, P. C. Myers: 'BIMA observations of infall in L1544: a young starless core in Taurus'. In: *The Physics and Chemistry of the Interstellar Medium, Proceedings of the 3rd Cologne-Zermatt Symposium, held in Zermatt, September 22-25, 1998* ed. by V. Ossenkopf, J. Stutzki, G. Winnewisser (GCA-Verlag Herdecke, 1999)
28. D. O. S Wood, E. Churchwell: ApJS **69**, 831 (1989)
29. Sh. Zhou, N. J. Evans, C. Koempe, C. M. Walmsley: ApJ **404**, 232 (1993)
30. H. Zinnecker, M. J. McCaughrean, B. A. Wilking: *Protostars and Planets III* (Univ. of Arizona Press), eds. E.G. Levy and J. Lunine, p. 429 (1993)

II. Disks, Jets and Protostars

2 A VLT/ISAAC Study of Circumstellar Disks and Envelopes Around YSOs

2.1 Abstract

We propose deep, high spatial resolution JHKL imaging of a nearby southern IRAS sample of 30 very young stellar objects (YSOs), selected from previous integrated near-infrared photometry to be faint and highly reddened, and also expected to be mostly nonstellar and nebulous. The main goal is to characterize and understand the morphology of the observed disk and envelope structures from the measured spatial surface brightness distribution. A few of these objects will turn out to have edge-on disks with dark dust lanes between extended scattering lobes. The VLT is needed to detect sub-arcsec structure in these nebulous young objects, including the wavelength-dependent sub-arcsec scaleheight of edge-on disks and hidden sub-arcsec infrared companions. In good seeing, ISAAC/VLT will deliver the most detailed YSO images ever seen, rivaling those of NICMOS/HST.

2.2 Scientific Rationale

The coplanar orbits of our planetary system suggest that a disk-like cloud surrounding the young Sun provided the raw materials for planet formation 4.6 billion years ago. Studies of nearby young stellar objects (YSOs) offer an opportunity to detect and characterize the disk environments which may play host to planet formation. Evidence has mounted that low-mass, pre-main sequence stars possess circumstellar disks with sizes and masses comparable to the early solar nebula. Unresolved infrared and millimeter continuum excess emission consistent with disk models is associated with more than half of the T Tauri stars in the Taurus-Auriga star-forming region [1]. Millimeter interferometers have mapped flattened molecular gas structures at 1 arcsec resolution around several T Tauri stars [3]. In some cases (*e.g.* GM Aurigae), evidence for Keplerian orbital motion in a circumstellar disk has been found (Dutrey, priv. commun.). Thus the existence of circumstellar (and possibly protoplanetary) disks around young stars is now well established.

The clearest images of circumstellar disks produced to date are 0.1 arcsec resolution optical images from the Hubble Space Telescope. As of this writing, six young stars associated with nearby molecular clouds show disks in reflected light. One is GM Aurigae [11], where the disk is seen adjacent to the bright point spread function of the central star. In the five other cases (HH30 [2]; Haro 6-5B [4]; DG Tau B and IRAS 04302 [9]; and HK Tau/c [12]), nearly edge-on disks are seen as absorption lanes bisecting the reflection nebulosities. In addition to these local young stars, seven backlit absorption disks have been detected toward the more distant Orion Nebula [7]. Taken together, all these HST results have provided some of the best available constraints on the structure and extent of young disks which may form solar systems like our own.

Disks viewed edge-on offer the best opportunity to study their internal structure, as the bright central star is occulted from view and thus does not pose a contrast problem. Furthermore, in edge-on systems the disk thickness can be directly measured. This vertical structure is related to the local disk temperature, and may also be affected by grain growth processes that are expected to initiate dust settling toward the disk mid-plane. In two of the edge-on disks studied so far with HST, HH 30 and HK Tau/c, detailed comparison of scattering models with the images has allowed disk scale heights to be measured for the first time [2] [12]. The disks' inclinations and masses have also been well-determined in these analyses, and the radial density profiles and flaring exponents have been constrained. Interesting differences are already apparent: the scale height of HH 30 is significantly larger than that of HK Tau/c, possibly due to the increased contribution of viscous heating to the energy budget. Resolved disks in young binary systems also present interesting tests for tidal perturbation theories. It is clear that edge-on systems provide most valuable laboratories for the study of disk structure and evolution. To understand the diversity of disk properties and the contexts in which they occur, it would be very beneficial to identify and study many more of these edge-on or nearly edge-on systems.

Two criteria are available to select young stars with edge-on disks. If disks and jets are perpendicular, as expected theoretically, then systems with low radial velocity jets should have nearly edge-on disks at the sources. The usefulness of this criterion has been demonstrated by the well-known imaging results for HH 30. However, all of the nearby YSO with published data indicating low radial velocity jets have been observed with HST already; and edge-on disks lacking bright jets (objects such as HK Tau/c) would not be identified by this criterion. Fortunately, there is a second powerful selection criterion: All of the edge-on YSO disks currently known distinguish themselves as unusually faint near-infrared sources as well as a large J-K colour index (> 4), indicative of high extinction (*cf.* Fig. 2 from [9]; see also [5] calling the object shown in Fig. 2 the "Butterfly Nebula"; however, they could not see the dark dust lane from their ground-based JHK observations, the dust lane only became apparent from the high-spatial resolution NICMOS JHK images. The images we propose to take with the VLT/ISAAC almost rival the spatial resolution of HST/NICMOS and should be good enough to resolve similar edge-on disks!)

This VLT/ISAAC proposal is based on the above photometric selection criterion; in fact we have collected the best possible southern object list for disk/envelope systems (faint, highly-reddened near-IR sources) a total of 30 objects (in the Cham, Rho Oph, Serpens, R CrA clouds). They are all so-called Class I objects, probably less than 1 Myr old. Our sample is 5 times

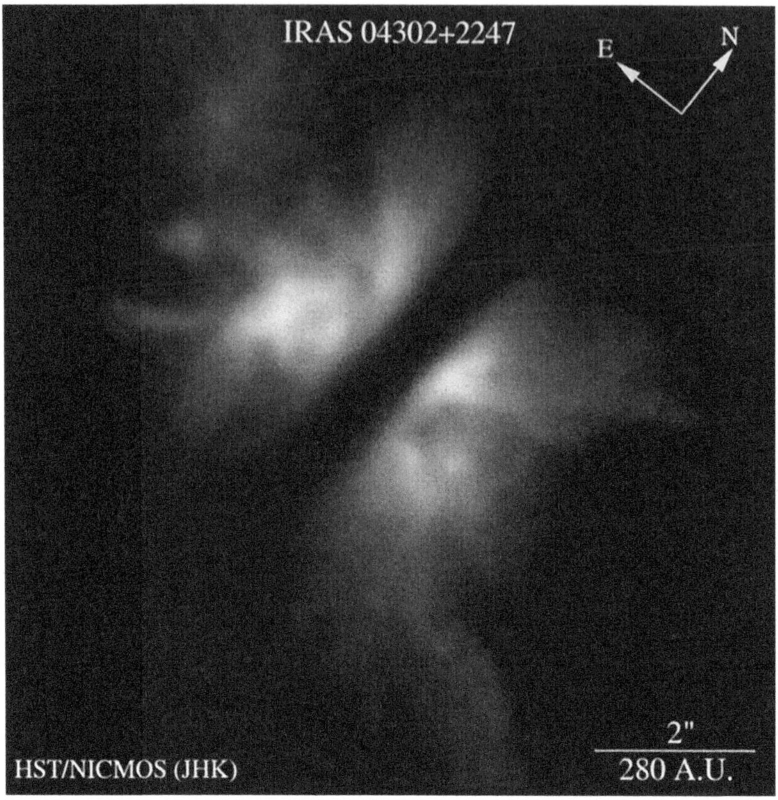

Fig. 2. Example of a disk/envelope system observed with NICMOS/HST

larger than the known disk sample from HST and promises to let us see the full diversity of disk/envelope properties, possibly leading to a systematic morphological classification or even an evolutionary sequence. For random orientation of the disks in space, we expect to find edge-on disks in about 10 % of the objects, which is why we need a large sample.

2.3 Immediate Objective

1) We will measure the surface brightness distribution of very young stellar objects still surrounded by tenuous envelopes left over from the star formation process. Based on their youth, they should have relatively large envelopes. These objects are younger than normal T Tauri stars and have not been observed systematically as a group yet. We will measure the surface brightness in several wavelengths to study the scattering morphology and the dust properties giving rise to the observed appearances and the observed nebular colours. For example, the high spatial resolution images of ISAAC will allow us to check if the scattering envelope is clumpy or smooth. We will provide an atlas/classification of these extended objects.

2) Eventually we will simulate the multi-wavelength appearance through radiative transfer model calculation (see [10]). Matching theory and observations, we will provide "initial conditions" for 3D hydrodynamical model calculations of continuing envelope accretion and disk formation. An evolutionary sequence might emerge, refining the Class I definition currently based soley on the integrated, spatially unresolved spectral energy distribution.

3) We expect to catch new examples of edge-on circumstellar disks. The JHKL observations can be used to infer a minimum mass for the disk from the estimated line-of-sight extinction (to be compared with the minimum solar mass nebula of $0.02\,M_\odot$). Images in different wavelengths in edge-on systems are also important to derive vertical scaleheights of circumstellar disks as a function of wavelength (all these issues are discussed in [8]).

4) We expect a number of other serendipitous discoveries in carrying out this YSO-VLT mini-survey. For example, we may discover new infrared companion, as most young stars have sub-arcsec companions [6], some of them very cool (large K-L, as in T Tau S), and possibly extended. Furthermore, our Class I sources may well be in an active phase, with outflow cavities discernible in scattered light and faint jets revealed in the K-band $2.12\,\mu$m molecular hydrogen line, while the central star itself is too heavily embedded to be seen (*cf.* HH212: [13]).

2.4 Telescope Justification

The need for deep, high spatial resolution (0.35 arcsec) infrared imaging calls for ISAAC on VLT/UT1. The only alternative would be NICMOS on HST, but this facility is no longer available (however, NICMOS would not have allowed us to take L-band images). The average image quality at the NTT ($\approx$0.7 arcsec) is not sufficient; furthermore there is no thermal infrared camera for L-band imaging at the NTT. Adaptive optics imaging with ADONIS at the ESO 3.6 m telescope is not possible for our sources, as the objects are too heavily obscured to serve as optical reference stars for wavefront sensing.

2.5 Results

Direct, near-infrared continuum observations at the VLT have helped us to find new examples of star-disk systems viewed nearly edge-on. These quasi edge-on disks are seen as spatially resolved dark lanes bisecting reflection nebulosity above and below the disk, with the nebulosity originating from the illumination of the dusty envelope or outflow cavities by the central star. A particularly nice case is the Chamaeleon infrared nebula (Fig. 3) in which the whole original pre-stellar cloud with 0.1 pc size shows up in scattered light. Presumably, these new star-disk systems resemble our own solar system in the early days of its formation (Figs. 4, 5).

Fig. 3. The Chameleon infrared reflection nebula, as seen with the VLT infrared camera ISAAC. The size of the image is $\sim 3 \times 2$ arcmin or $\sim 0.15 \times 0.10$ pc at a distance of 140 pc.

References

1. S. V. W. Beckwith, A. I. Sargent, R. S. Chini, R. Guesten: AJ **99**, 924 (1990)
2. Ch. J. Burrows, K. R. Stapelfeldt, A. M. Watson et al.: ApJ **473**, 437 (1996)
3. A. Dutrey, S. Guilloteau, G. Duvert et al.: A&A **309**, 493 (1996)
4. J. E. Krist, K. R. Stapelfeldt, Ch. J. Burrows et al.: ApJ **501**, 841 (1998)
5. P. W. Lucas, P. F. Roche: MNRAS **286**, 895 (1997)
6. R. D. Mathieu: ARAA **32**, 465 (1994)
7. M. J. McCaughrean & C. R. O'Dell: AJ **111**, 1977 (1996)
8. M. J. McCaughrean, K. R. Stapelfeldt, L. M. Close: 'High-Resolution Optical and Near-Infrared Imaging of Young Circumstellar Disks'. In: *Protostars and Planets IV*, ed. by A. P. Boss, S. S. Russell (University of Arizona Press, Tucson 2000) p. 485
9. D. Padgett, W. Brandner, K. R. Stapelfeldt et al.: AJ **117**, 1490 (1999)

Fig. 4. VLT / UT1 - ISAAC JHK observations of four young stellar objects in the Gum Nebula. The objects are HH 46 / 47 (upper left), CG 30 (upper right), Re 4 (lower left), and Re 5 (lower right). North is up and east is left for HH 46 / 47, CG 30, and Re 4; Re 5 has north right and east up. The field sizes are (in arcsec): 143 × 123, 121 × 130, 86 × 143, 125 × 140, respectively. All objects are seen to show outflow activity. (See: H. Zinnecker, A. Krabbe, M. J. McCaughrean et al.: 'A search for young solar system analogues with the VLT', A&A **352**, L73, (1999))

10. S. Richling, H. W. Yorke: A&A **327**, 317 (1997)
11. K. Stapelfeldt, Ch. J. Burrows, J. E. Krist: 'Hubble Space Telescope Imaging of the Disks and Jets of Taurus Young Stellar Objects'. In: *Herbig-Haro Flows and the Birth of Stars; IAU Symposium No. 182*, ed. by B. Reipurth, C. Bertout (Kluwer Academic Publishers 1997)
12. K. R. Stapelfeldt, J. E. Krist, F. Menard et al.: ApJ **502**, L65 (1998)
13. H. Zinnecker, M. J. McCaughrean, J. T. Rayner: Nature **394**, 862 (1998)

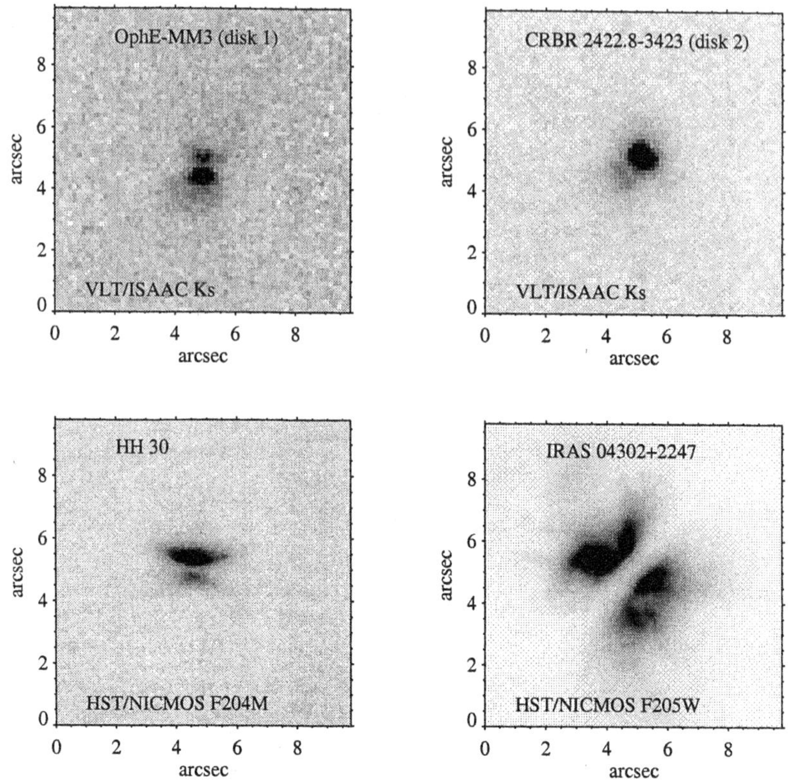

Fig. 5. Edge-on circumstellar disk sources in Ophiuchus and Taurus. In comparison to isolated disks in Taurus, the Ophiuchus disks and their reflection nebulosities are more compact. For the disk sources in the Ophiuchus region, North is up and East is to the left. (See: W. Brandner, S. Sheppard, H. Zinnecker et al.: 'VLT-detection of two edge-on Circumstellar Disks in the rho Oph dark cloud', A&A **364**, 13 (2000))

3 Infrared H_2 Imaging of the Highly Symmetric Proto-stellar Jets HH 212 and HH 111

3.1 Abstract

We propose to obtain deep diffraction-limited images of two highly collimated protostellar jets in the v = 1 − 0 S(1) line of molecular hydrogen at 2.122 μm using NICMOS. HH 111 is a well-known jet visible at optical wavelengths, and the new NICMOS data will be analyzed in conjunction with existing WFPC2 imaging. HH 212 is a newly discovered jet, obscured by its parent molecular cloud core and only detectable at near-IR wavelengths. The high degree of symmetry seen in both jets makes them crucial template objects for studying the earliest phases of jet formation, since the underlying "patterns" have not been erased by strong environmental effects. The new images will however enable us to study the interaction of the jets with their surroundings, where the H_2 line traces shocks at the interface between the jet and the ambient gas. In particular, we will examine the hypothesis that molecular outflows are created by the entrainment of gas into the wings of jet bow-shocks. The Cycle 7 data will be used as the first epoch in proper motion study: by measuring transverse velocities as small as $40\,km\,s^{-1}$ over the NICMOS mission lifetime, we will study the dynamical evolution of the jets in comparison with theoretical models. Finally, we will use deep continuum images to determine the structure of faint reflection nebulae seen around the jet driving sources, in order to determine whether these are cavities excavated by the molecular outflows, or the upper surface of a large-scale disk around the source.

3.2 Scientific Background: Star Formation and Herbig-Haro Jets

Herbig-Haro (HH) jets from young stellar objects are ubiquitous; well over 300 flows are known [12], and they have been a subject of intense study over the last decade (see, *e.g.*, [14]). Highly collimated jets and associated outflows are observed from even the most deeply embedded protostars, and thus they must be an integral part of star formation and the earliest phases of stellar evolution. Unlike extragalactic jets, HH jets emit line radiation and are nearby enough that they are resolvable in ground- and space-based observations. Not only can their morphological structure be analyzed, but also their kinematics can be explored through a combination of radial velocity and proper motion studies. Thus, it is possible to derive detailed physical properties for these objects.

The jets typically consist of a highly collimated chain of knots terminating in a working surface where the flow hits the ambient molecular cloud. Typical jet velocities and dimensions are $\sim$ few $\times\ 100\,km\,s^{-1}$ and ~ 0.3 pc, suggesting dynamical ages of a few 10^3 yrs. The jets often have multiple working surfaces along their flow axes, understood in terms of episodic activity of the driving source, probably linked to massive accretion events in its circumstellar disk.

Most known HH jets emit optical emission lines ([S II] and Hα): only a much smaller number of infrared jets, traced in the $v = 1-0$ S(1) line of shock-excited H_2, are known to date. Yet these are important sources, since to study the earliest phases of stellar evolution, *e.g.*, the history of accretion/ejection phenomena, we need to study embedded jets from embedded protostellar objects. Of the known infrared jets, there are two which exhibit a very high degree of collimation and two-sided spatial symmetry, HH 212 and HH 111.

The Targets: HH 212 and HH 111

HH 212 — Discovered through imaging in the $2.12\,\mu$m line of H_2, HH 212 is totally embedded and not optically visible [27]. It is a flow of extraordinary bilateral symmetry (Fig. 6), with matching series of inner knots and outer bow shocks on both sides, extending over a total length of 240 arcsec

(0.6 pc in Orion). It is accompanied by a highly collimated molecular outflow extending the full length of the jet (Dent, personal communication). Radial velocities measured for the H_2 knots on either side of the centre are almost identical, indicating that HH 212 lies very close to the plane of the sky. Exactly between the two sides of the jet lies a cold, low-luminosity ($15\,L_{\odot}$) IRAS and mm-continuum source [26], coincident with a strong 1.3 cm water maser [24]. Millimetre interferometry reveals highly collimated shocked SiO along the inner jet, and a rotating $C^{18}O$ disk perpendicular the source [10]. Evidence for this disk and its extended envelope can also be seen in the near-IR continuum, where the invisible central source is straddled by a matching pair of paraboloidal reflection nebulae, reminiscent in structure (if not in scale) to those imaged around the HH 30 driving source with the HST [1].

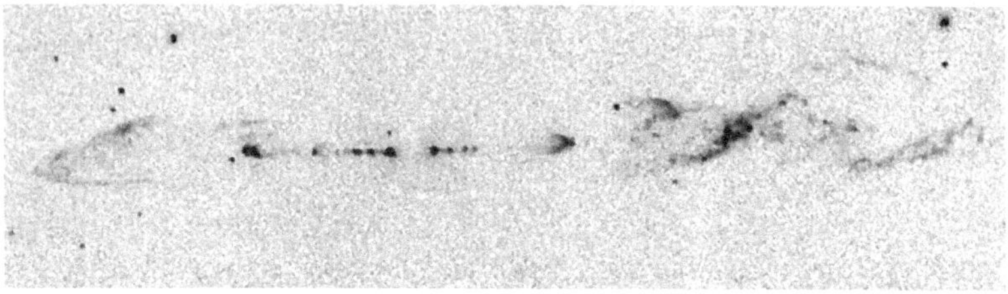

Fig. 6. Ground-based near-IR v = 1 − 0 S(1) H_2 mosaic image of HH 212. The image is ~ 250 arcsec across, ~ 0.5 pc in Orion. The spatial resolution is ~ 0.7 arcsec. The data were taken at the Calar Alto 3.5 m telescope. The extreme bilateral symmetry of the jet is apparent, both on small scales (inner knots) and large (outer bow shocks).

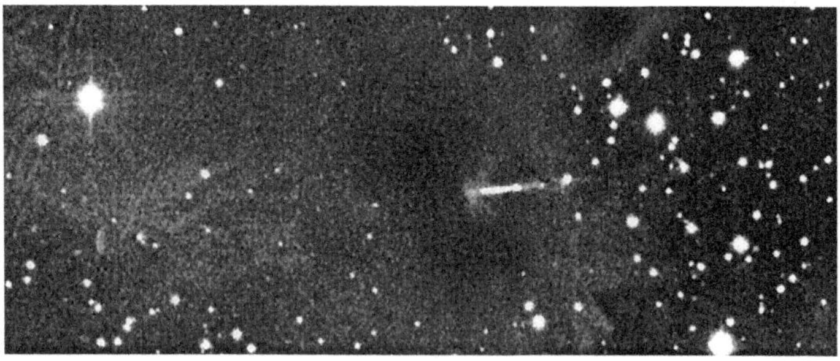

Fig. 7. Ground-based optical ([S II]) CCD mosaic image of HH 111. The image is ~ 500 arcsec across, ~ 1.1 pc in Orion. The bright knots and bow shocks in the west have faint symmetric counterparts in the east: this symmetry is considerably enhanced in the near-IR.

HH 111 — Discovered by Reipurth [11], this is one of the finest HH jets known. Fig. 7 shows a ground-based CCD mosaic of the HH 111 complex in Orion. The jet emerges from a dense cloud core, which harbours a 25 $L_\odot$ IR source. The source illuminates a conical reflection nebula through which the jet appears. After a series of knots, the flow terminates in a working surface 142 arcsec (0.32 pc) from the source. On the opposite side of the source, two other working surfaces are found at 200 and 226 arcsec. Overall the jet takes on a more symmetric two-sided appearance when imaged in H_2 [5] [6]. The jet complex is ~0.8 pc in length [11], and with the more recent discovery of even more distant working surfaces, Reipurth, Bally & Devine [18] find that in total the HH 111 complex extends more than 7 pc. A molecular outflow coincides with the jet [13], and mm-interferometric observations show the CO flow surrounding the optical jet as a hollow tube of molecular material, probably dragged along by entrainment processes [15]. Optical spectroscopy of the jet shows moderate radial velocities, with the western lobe approaching us, and the eastern lobe receding. Combined with measured large proper motions (300 – 600 $km\,s^{-1}$) away from the source, an inclination of the jet to the plane of the sky of ~10 degrees is derived [16]. We have observed the HH 111 jet with HST/WFPC2, obtaining high signal-to-noise Hα and [SII] images [19]. The inner jet is fully resolved in Hα (Fig. 8), and analysis shows that the jet body consists of a series of small internal working surfaces, which are also prominent in the near-IR H_2 images. The driving source of the HH 111 jet is deeply embedded, surrounded by large amounts of cold dust and gas [14] [23]. A rotating disk has been resolved, lying perpendicular to the jet axis [25]. The source is detected by the VLA at 3.6 cm, and high angular resolution (0.2 arcsec) maps reveal a tiny one-sided radio jet well aligned with the optical jet [20].

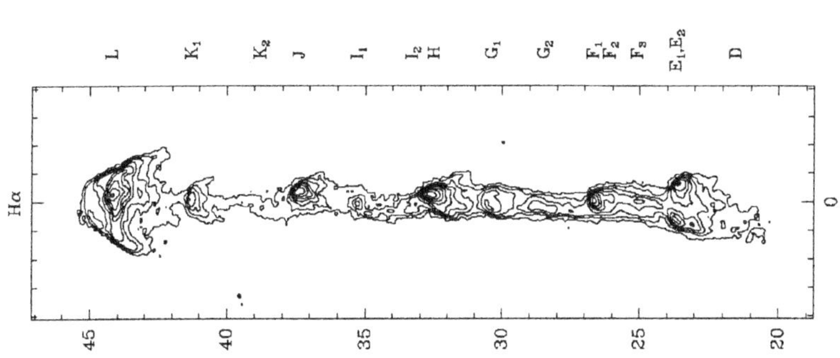

Fig. 8. Contour plot of the HST Hα image of HH 111. The jet is fully resolved and is seen to comprise of a series of internal working surfaces. The plot covers 30 arcsec along the brightest part of the optical jet.

HH 111 and HH 212 share many characteristics, including their extremely high collimation, the high degree of two-sided symmetry, the association with very collimated molecular outflows, and low-luminosity proto-stellar driving sources with near edge-on circumstellar disks. Together they represent the Herbig-Haro phenomenon at its best.

3.3 Scientific Goals

In the last few years, it has been realized that molecular hydrogen is a major coolant in Herbig-Haro jets. Near-IR array images of many outflow regions in addition to HH 111 and HH 212 have revealed substantial H_2 emission with a variety of geometries, *e.g.*, HH 211 [9], HH 91 [7], HH 47 [3] and Ceph A [8]. It is becoming increasingly clear that the rich morphological variety of H_2 emission is the key to understand evolutionary effects, flow types, and environmental aspects of the observed jets.

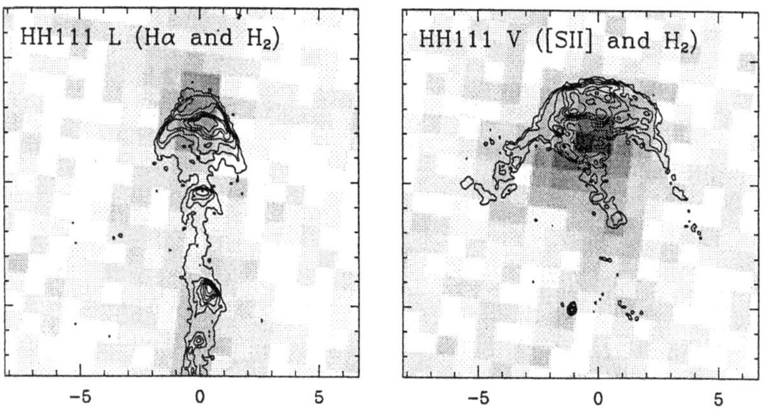

Fig. 9. Contour plots from the HST emission-line of two of the bright bow shocks in HH 111 overlaid on corresponding ground-based near-IR H_2 images. Bow shock L is shown in Hα on the left; bow shock V in [S II] on the right.

Jet Morphology: shocks and entrainment
As a jet propagates, it interacts vigorously with the ambient medium sweeping up and carrying along some of the surrounding gas. This entrainment can take place in the bow shock at the head of the jet (referred to as *prompt entrainment)*, or in a turbulent boundary layer along the sides of the jet (referred to as *steady entrainment*). These two processes should be observationally distinct. In the case of steady entrainment, models suggest that the H_2 emission should be brightest at the edges of the jet. In an extreme case of steady entrainment, the jet may dissipate and form a collimated flow of

turbulent molecular gas [22]. Conversely, in the case of prompt entrainment, the warm H_2 will be swept to the side by the wings of the bow shock, and a series of bow shocks should then form a shell of H_2 that surrounds the optically emitting gas [8].

Similarly, the morphology of the H_2 emission can tell us about the type of shock involved. When a shock proceeds into mostly neutral gas, and if a strong magnetic field is present, then a "C" shock forms [2], and the H_2 should coincide with the wings of the optical emission [21]. On the other hand, models of slow "J" shocks suggest the H_2 emission should occur just inside the optical bow shock wings. If fluorescence plays a role in exciting H_2, we would expect the H_2 to radiate in a diffuse photo-dissociation region surrounding the bow shock (*e.g.*, [4]). As a consequence, the detailed morphology of H_2 can provide significant insights into the nature of the flow.

In the case of the optically visible, approaching lobe of the HH 111 jet, we can directly compare the locations of the optical and molecular emission. The Hα line, in particular, serves to trace the precise position of the shock front, while [S II] emission marks the more extended post-shock cooling zone. Comparing our optical HST images of HH 111 with ground based H_2 data (Fig. 9) provides tantalizing hints, but is inconclusive, because of the low resolution ($\sim$1 arcsec) of the ground-based H_2 frames. For example in HH 111 L, the working surface at the tip of the bright jet, the H_2 emission is strongest at the apex of the bow shock, but it is impossible to say whether the emission peaks at the shock (C-shock) or slightly behind it (J-shock). In the outer working surface HH 111 V, the H_2 emission clearly peaks well behind the bow shock, but in this case more probably arises in the Mach disk or reverse shock which decelerates the jet. Faint H_2 emission can be seen along the edges of the jet in Fig. 9, but given the poor resolution one cannot judge whether this comes from a continuous boundary layer or from the extended wings of the various bow shocks. A comparison of the existing optical HST data with similar resolution H_2 images would evidently provide an essential advance in our understanding of shock processes in this jet.

HH 111 is a case of an optically visible jet which has been excavated partly by its own interaction with its surroundings and partly because the OB stars in the Orion Ib association have helped sweep away the surrounding gas. HH 212, on the other hand, is a pure infrared jet, and is likely to be in an even earlier evolutionary stage than HH 111: it is deeply embedded; it has a much smaller total extent than HH 111; its source is even colder than the HH 111 source. HH 212 is the *only* infrared jet known which has the perfect collimation seen in the finest optical jets: all other infrared "flows" are surrounded by significant H_2 envelopes or clumps, which complicate study of the jets themselves. The remarkable symmetry in geometry and brightness of the two lobes of HH 212 makes it ideal for a study of jet structure, and a comparison with the large number of jet models discussed above.

By comparing the HH 212 and HH 111 jets at high spatial resolution, we can hope to answer the question of whether the differing environment and evolutionary state of the two jets affects the knot structure or if they are morphologically identical despite these factors.

Proper Motions: Kinematic Structure and Dynamical Evolution
We also wish to study proper motions in the jets, particularly in HH 212, in order to understand the kinematics of the detailed H_2 structures. The present data will provide a first epoch which we would then follow-up later in the NICMOS mission. The expected bulk jet velocity is on the order of $200-500\,km\,s^{-1}$, which, since the jet lies close to the plane of the sky, translates into transverse proper motions of 0.1 – 0.25 arcsec/year at 450 pc. Such proper motions should be well measured over the lifetime of NICMOS: over a 4 year baseline, $40\,km\,s^{-1}$ is equivalent to 0.1 arcsec, or half the diffraction-limit. Therefore, we will request observations in Cycle 10/11 similar to those proposed here, assuming they are successful.

Reflection Nebulae: Outflow Cavity or Circumstellar Envelope?
Finally, we wish to study the faint, diffuse continuum nebulae seen at the base of both jets: in HH 212, again a symmetric pair is seen either side of the driving source. We assume these to be reflection nebulae illuminated by the driving source, but what is not clear, is how the characteristic paraboloidal shape arises. There are two basic models. First, the reflection nebula may trace a cavity around the jet excavated by an associated less collimated wind. Second, the nebulae may trace the upper (and lower) surface of the circumstellar disk around the driving source, as beautifully seen in the HST images of the HH 30 jet [1]. However, the reflection nebulae at the base of both HH 111 and HH 212 are much larger and further displaced from the driving source, and they may trace a larger-scale infalling envelope rather than the centrifugally-supported disk. We wish to study the shape of these reflection nebulae in detail, in order to distinguish between the two models, and provide constraints on the flaring of either the outflow cavity (*e.g.*, how collimated is the wind compared to the jet?) or the circumstellar envelope (*e.g.*, what is the envelope aspect ratio?). To do so, we propose taking very deep broad-band F160W NIC3 images of the reflection nebulae taking advantage of the very low sky background at $1.6\,\mu m$. High spatial resolution is needed in order to trace the nebular shapes as accurately as possible.

3.4 The Need for HST

The highest possible spatial resolution is needed in order to resolve the smallest-scale structures in the jets: typical H_2 cooling lengths for gas at 2000 K and $10^5\,cm^{-3}$ are $10^{15}\,cm$ [21], *i.e.*, 66 AU or 0.15 arcsec at 450 pc. This is just below the HST diffraction limit at $2.122\,\mu m$.

To study the jet kinematics, we need observations with HST at two epochs. To achieve a *velocity* resolution of $\sim 40\,km\,s^{-1}$ with ground-based observations would require $\sim$20 years. However, with the poor *spatial* resolution of ground-based data, they can only provide the bulk motion of the flow, while it is the kinematical behaviour of the fine structural details that are important, since that outlines the dynamical processes at work in the jets. Moreover, we expect significant changes in the knot structures (fading, merging, *etc.*) on these longer time-scales: therefore it is important to measure the velocities over the shortest possible interval, to separate the effects of material motion and excitation changes. Thus, short timescale, high resolution observations are required, thus the need for HST.

Ground-based adaptive optics will *not* work for either HH 212 or HH 111, since there are no stars brighter than $m_v \sim 17^m$ within 1 arcmin of either jet.

Acknowledgments. This HST proposal was written by M. J. McCaughrean (PI) and B. Reipurth, with HZ as a co-investigator.

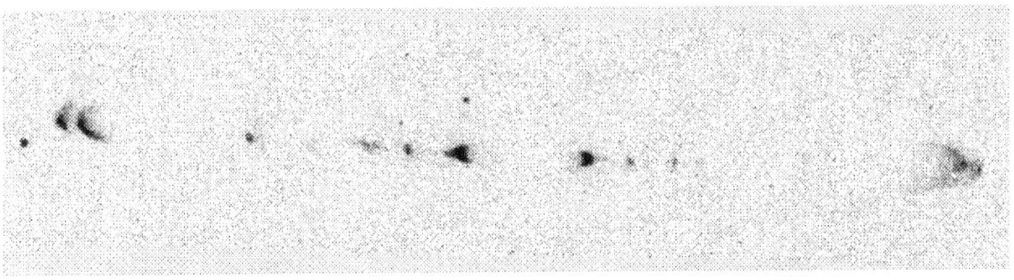

Fig. 10. A closer look at the innermost part of the twin-exhaust H_2 jet HH 212 based on diffraction-limited $\lambda = 2.12\,\mu$m imaging with HST/NICMOS2 (*cf.* Fig. 6). The mini bow-shocks that are revealed are not completely mirror-symmetric. Note the interknot H_2 emission.

3.5 Results

Diffraction-limited, near-infrared emission line observations of the highly symmetric, pulsed, twin-exhaust jet HH 212 have been made in molecular hydrogen with NICMOS on HST ($\lambda = 2.12\,\mu$m, 0.25 arcsec resolution) in order to take a closer look at the detailed spatial structure of the innermost collisionally excited H_2 knots nearest to the invisible exciting source (see Fig. 10). The knots are seen to be mini bow-shocks but their bilateral mirror symmetry is less striking than in previous images of poorer resolution, implying that either the gaseous medium into which the jet is propagating is slightly asymmetric (*cf.* Wiseman et al. 2001, ApJ 550, L87) or that the jet driving mechanism itself is somewhat asymmetric with respect to the direction of ejection. Interknot emission is also revealed, indicating weak shock heating of the ambient gas all along the inner jet channel. Note that in the case of a centrally pulsed jet, jet gas ejected recently at high speed will be propagating through a partially empty channel, interacting with the channel walls, and finally catching up and running into the jet gas ejected earlier, perhaps creating both the interknot emission and the shock-heated knots. (The situation as a whole is reminiscent of a series of car accidents on a motorway with groups of cars driving at different speeds, skidding along the crash barrier and ultimately crashing into each other!).

The same HST experiment on a second jet (HH 111) did not add any new insight, because its H_2 emission was too faint to be seen.

PS. Another surprising result on HH 212 is the following: comparison of the H_2 emission line and 2 μm continuum images suggests that the reflection nebulae, which are also seen near the inferred location of the exciting source, are *not* due to reflected 2 μm continuum starlight, but most likely due to reflected H_2 line emission, with the innermost H_2 knots bright enough to provide the dust illumination.

References

1. Ch. J. Burrows, K. R. Stapelfeldt, A. M. Watson et al.: ApJ **473**, 437 (1996)
2. B. Draine: ApJ **241**, 1021 (1980)
3. J. Eislöffel, C. J. Davis, T. P. Ray, R. Mundt: ApJ **422**, L91 (1994)
4. A. Fernandes, P. W. J. L. Brand: MNRAS **274**, 639 (1995)
5. R. Gredel, B. Reipurth: ApJ, **407**, L29 (1993)
6. R. Gredel, B. Reipurth: A&A, **289**, L19 (1994)
7. R. Gredel, B. Reipurth, S. Heathcote: A&A, **266**, 439 (1992)
8. P. Hartigan, J. Carpenter, C. Dougados, M. F. Skrutskie: AJ, **111**, 1278 (1996)
9. M. J. McCaughrean, J. T. Rayner, H. Zinnecker: ApJ, **436**, L189 (1994)
10. M. J. McCaughrean, A. I. Sargent: in preparation (1997)
11. B. Reipurth: Nature, **340**, 42 (1989)
12. B. Reipurth: 'A General Catalogue of Herbig-Haro Objects', *Electronically published via anon. ftp to ftp.hq.eso.org, directory pub/Catalogs/Herbig-Haro*, (1994)
13. B. Reipurth, M. Olberg: A&A, **246**, 535 (1991)
14. B. Reipurth, S. Heathcote: In: *STScI Symposium Astrophysical Jets*, ed. by D. Burgarella, M. Livio, C. P. O'Dea (Cambridge University Press, Cambridge 1993) p. 35
15. B. Reipurth, J. Cernicharo: 'Herbig-Haro Jets at Optical, Infrared and Millimeter Wavelengths'. In: *Revista Mexicana de Astronomia y Astrofisica Serie de Conferencias, Vol. 1, Circumstellar Disks, Outflows and Star Formation, Cozumel, Mexico, Nov 28-Dec 2, 1994*, ed. by S. Lizano, J. M. Torrelles, (1995) p. 43
16. B. Reipurth, A. C. Raga, S. Heathcote: ApJ, **392**, 145 (1992)
17. B. Reipurth, R. Chini, E. Krügel, E. Kreysa, A. Sievers: A&A, **273**, 221 (1993)
18. B. Reipurth, J. Bally, D. Devine: AJ **114**, 2708 (1997)
19. B. Reipurth, P. Hartigan, S. Heathcote, J. Morse, J. Bally: AJ, **114**, 757 (1997)
20. L. F. Rodriguez, B. Reipurth: A&A, **281**, 882 (1994)

21. M. D. Smith, P. W. J. L. Brand: MNRAS, **245**, 108 (1990)
22. S. Stahler: ApJ, **422**, 616 (1994)
23. K. R. Stapelfeldt, N. Z. Scoville: ApJ, **408**, 239 (1993)
24. J. G. A. Wouterloot, C. M. Walmsley: A&A, **168**, 237 (1986)
25. J. Yang, N. Ohashi, J. Yan, C. Liu, N. Kaifu, H. Kimura: ApJ, **475**, 683 (1997)
26. H. Zinnecker, P. Bastien, J. P. Acoragi, H. W. Yorke: ApJ, **265**, 726 (1992)
27. H. Zinnecker, M. J. McCaughrean, J. T. Rayner: 'HH 212: The Most Beautiful Protostellar Jet Known to Date'. In: *Disks and Outflows Around Young Stars, Proceedings of a Conference Held at Heidelberg, Germany 6 - 9 September 1994*, ed. by S. V. W. Beckwith, J. Staude, A. Quetz, A. Natta, (Springer-Verlag Berlin Heidelberg New York 1996), Also: Lecture Notes in Physics **465**, 236 (1996)

The reader is also referred to the following, more recent papers:
H. Zinnecker, M. J. McCaughrean, J. T. Rayner: 'A symmetrically pulsed jet of gas from an invisible protostar in Orion', Nature **394**, 862 (1998) (HH 212)

B. Reipurth: 'Disintegrating Multiple Systems in Early Stellar Evolution', AJ **120**, 3177 (2000) (HH 111)

4 An Unbiased mm Continuum Survey for Protostellar Outflow Sources in the Orion A Cloud

4.1 Abstract

We propose to use the MPIfR 37-channel bolometer array at the IRAM 30 m telescope to carry out for the first time an *unbiased* survey for 1.3 mm continuum emission from cool dust in protostellar objects over a large part of the Orion A cloud. Our goal is to map a total area of ~ 1 square degree in an effort to find the sources of embedded jets found in our unbiased infrared H_2 survey of the same area. The millimetre survey will very effectively reveal even the youngest (Class 0) objects. It is also important to establish the population of sources *without* corresponding jets. Our combined infrared/mm-survey constitutes the first statistical investigation addressing the earliest phases of protostellar jets and their feedback on the scale of a full molecular cloud.

4.2 The Need for an Unbiased Millimetre Survey for Protostellar Outflow Sources

Once a prestellar cloud core becomes unstable to gravitational collapse, a young star is formed at its centre surrounded by a dusty envelope and a circumstellar accretion disk (see, *e.g.* [1]; [9]). Since dust effectively absorbs shorter wavelength radiation, the youngest objects (Class 0 protostars), surrounded by the most massive dust envelopes, can only be found by detecting the submillimetre and millimetre wavelength emission from cool dust in the disk / envelope. Also, at a very early evolutionary stage, protostars are found to drive energetic outflows, probably due to the presence of accretion

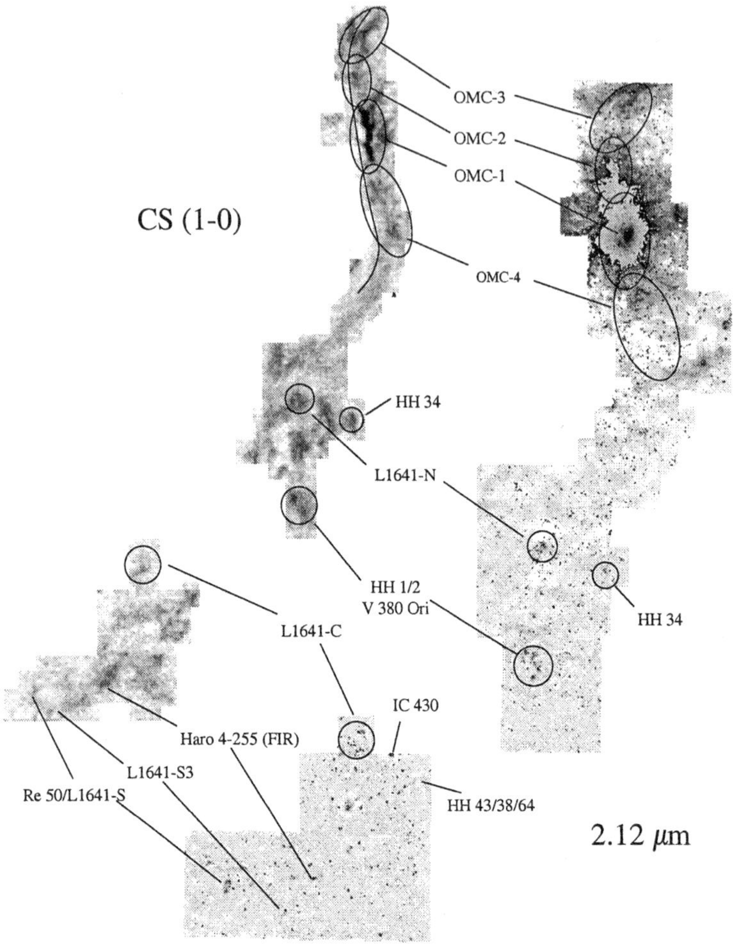

Fig. 11. The figure shows on the left panel the distribution of dense molecular gas as traced by the CS(1-0) transition [8]; the frame around the map is about 2.5×1.5 degrees wide). There are 125 dense cloud cores, thought to be the locations of ongoing or future star formation. The CS-map also delineates the area that we propose to survey at 1.3 mm. The right panel shows the current state of the H_2 survey (the image shows a drastically compressed version of the big mosaic, which has 0.4 arcsec pixels). Based on the number of H_2-jets found in that big mosaic, we expect to detect a total of about 40 jets in the final image. This means that by far not all of the dense molecular clumps identified in the CS-maps are associated with H_2-jets, either indicating that these currently do not form stars or that there is star formation without jets. The latter possibility can be tested with the proposed millimetre survey, which should detect all protostellar objects. That's why we must map the *whole* region (*unbiased mm-survey*). Credit: Th. Stanke (Ph.D. Thesis)

through the disk. We are currently conducting an unbiased infrared survey in the Orion A cloud, designed to detect emission from shock-excited H_2 in the flows ([7]; see also Fig. 11). *Here we propose to carry out an unbiased millimetre survey of the same region for the driving sources of these outflows.* Subsequent follow up FIR / submm-bolometry will allow us to determine the nature and evolutionary status of these driving sources, in particular whether they are Class I or Class 0 sources. This will enable us to establish on a statistical basis the moment at which protostellar jets first develop. The basic reason why we need an *unbiased* survey of the *whole* cloud at mm wavelengths is to reveal the full population of protostellar cores and clumps, and in particular to investigate what fraction of this population may be without any sign of outflow activity.

4.3 Framework: A Survey for Jets and Protostars in L 1641

The Jet-Disk Connection

Over the past decade, it has been established that accreting YSOs reveal themselves via simultaneous mass loss in the form of high-velocity molecular outflows and/or highly collimated jets. Central magnetised accretion disks are thought to launch the jets and outflows in protostellar systems, as in active galactic nuclei [2]; [5]; [3], and theoretically at least, there appears to be a strict jet-disk connection, *i.e.* no jet without an accretion disk, and no accretion disk without a jet. Therefore, we can expect jets and outflows to be seen at the very earliest stages of protostellar evolution, *i.e.* from Class 0 objects. Thus by searching for very young jets and outflows, we can hope to identify the protostellar jet driving sources, as it has been done in the case of HH 212 [11]; [12] and recently for V 380 Ori NE (see Fig. 12).

The Survey Region. The Orion molecular cloud complex at 450 pc is the nearest site of ongoing high- and low-mass star formation. Tatematsu et al. [8] surveyed a ~ 1 square degree region of the Orion A cloud in CS (1-0), finding 125 dense molecular gas clumps, thought to be the locations of ongoing or future star formation. The large variety of environments present in the cloud (high mass star formation in the BN / KL complex; clustered and distributed low- to intermediate-mass star formation all over the cloud) ensures that surveys in this region will provide samples of objects largely free from selection effects.

The Jet Survey. To search for the *youngest, embedded* flows in Orion A we make use of the emission from shock-excited molecular hydrogen at $\lambda = 2.12\,\mu$m, which, with the advent of large format IR-cameras, is much more efficient than using millimetre spectroscopy to search for high-velocity molecular gas. The survey, using the Calar Alto 3.5 m telescope and a 1024×1024

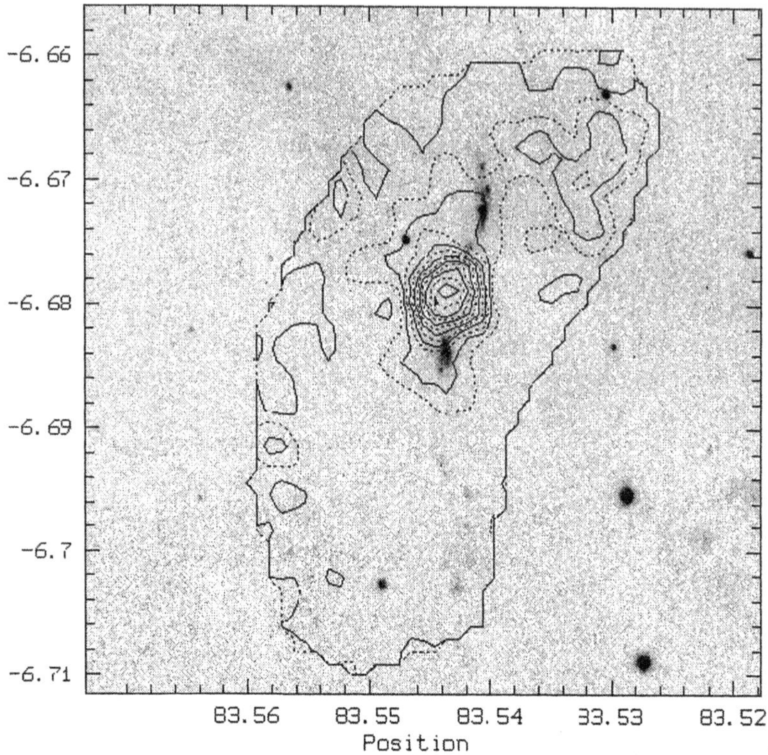

Fig. 12. This image shows the result of a test observation (data taken by Karl Menten during January 1998 with the 37-channel bolometer). The image is an overlay of an infrared image highlighting molecular hydrogen emission features (grayscale) and a 1.3 mm map (contours). We find that the mm-source is located right on the jet axis. Since there is no obvious point source in the IRAS images, the mm-source probably is a true Class 0 protostellar object, "found" by near-infrared imaging (plus follow-up mm imaging), demonstrating the power of our method.

pixel IR camera, is now almost complete. The analysis of ~20 % of the data indicates the power of this approach, with ~10 new jet-like sources already discovered ([7]; see also Fig. 11). A complementary unbiased optical survey for giant Herbig-Haro-flows, scheduled for January 1999 at ESO, will reveal the true extent of the flows and the role of interactions of *more evolved flows* with the cloud on larger scales.

The Jet Sources. Accompanying near-/mid-IR observations at *J, H, K* and *N* will also help find and characterize the outflow sources. However, the true test cases for the jet-disk connection are the youngest sources, just turning from clumps of gas into star+disk systems. Since these are detectable only at long wavelengths ($\geq 100\,\mu$m), we require an equivalent unbiased millimetre continuum survey to probe the jet-disk connection even at these youngest stages, which are most critical for the onset of outflows.

4.4 Technical Details

We propose to use the MPIfR 37-channel bolometer array at the 30m telescope to survey a total area of $\sim$1 square degree in the Orion A cloud for 1.3 mm dust continuum emission, as a probe for very young protostellar objects. This instrument is well suited for this task because of its high sensitivity and the possibility of mapping large areas with good angular resolution (of the order of 10 arcsec) in a reasonable amount of time.

Very young stellar objects in Orion, like the mm-condensations in OMC2 and 3 [4] and some well known HH energy sources [6], have 1.3 mm fluxes of typically a few hundred mJy. For the survey proposed here we suggest to try to reach a detection limit of about 30 mJy rms, in order to detect sources of that kind and possibly younger, fainter ones. Judging from pre-stellar cores in Taurus (1.3 mm: a few tens of mJys at 150 pc; [10]) we would expect similar cores in Orion to be at a few mJy and thus below our detection limit; however, in Orion the pre-stellar cores may be more massive and warmer, thus we might well detect them in our survey. Since the transition from pre-stellar to protostellar objects is (by definition) the point at which accretion starts (at which point outflow activity is also expected to start), we must try to detect the youngest and thus faintest Class 0 objects. Therefore we propose to reach a sensitivity limit of about 60 – 120 mJy at 1.3 mm (*i.e.* an rms noise of $\sim$20 - 40 mJy), in order to find these youngest protostars.

4.5 Results

The 1 square degree unbiased IRAM 1.3 mm continuum survey to find embedded jet driving sources associated with dense protostellar dust condensations produced a number of detections. In addition, a few cold dust condensations were discovered *without* molecular jets; apparently these dense condensations have not formed a star yet, but are likely to do so in the future. The small number ratio of pre-stellar and stellar (jet-related) condensations may be taken to indicate that the onset of star formation in the Orion A cloud was partly synchronised by an external triggering event (Stanke et al., in prep.); see also: Th. Stanke, M. J. McCaughrean, H. Zinnecker: 'Giant protostellar outflows revealed by infrared imaging', A&A **355**, 639 (2000)).

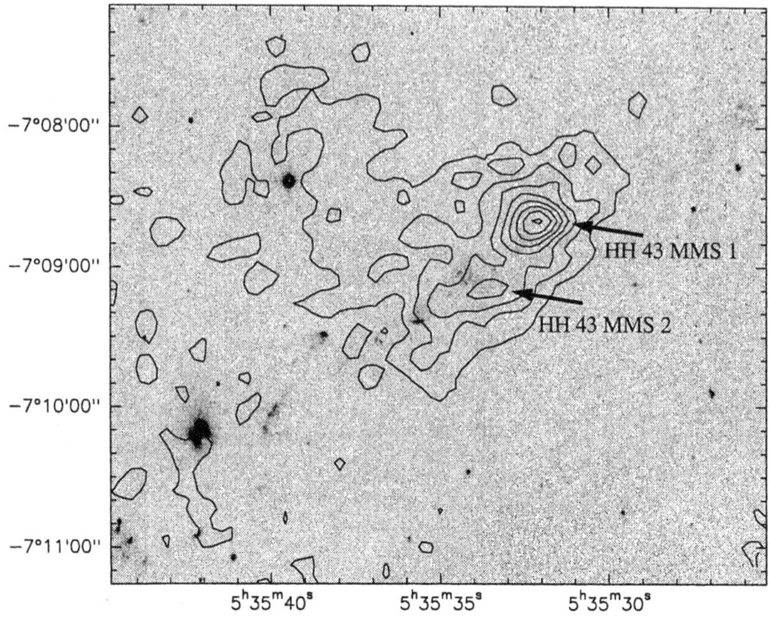

Fig. 13. Example of a successful search for a protostellar cold dust continuum source (1.3 mm) associated with a Herbig-Haro flow (jet driving source).

References

1. P. André, D. Ward-Thompson, M. Barsony: ApJ, **406**, 122 (1993)
2. R. D. Blandford, D. G. Payne: MNRAS, **199**, 883 (1982)
3. M. Camenzind: 'Energetics, Collimation and Propagation of Galactic Protostellar Outflows. Views and Perspectives'. In: *Herbig-Haro Flows and the Birth of Stars; IAU Symposium No. 182*, ed. by B. Reipurth, C. Bertout (Kluwer Academic Publishers 1997)
4. R. Chini, B. Reipurth, D. Ward-Thompson et al.: ApJ, **474**, L135 (1997)
5. R. E. Pudritz, C. A. Norman: ApJ, **301**, 571 (1986)
6. B. Reipurth, R. Chini, E. Krügel et al.: A&A, **273**, 221 (1993)
7. T. Stanke, M. J. McCaughrean, H. Zinnecker: A&A, **332**, 307 (1998)
8. K. Tatematsu, T. Umemoto, O. Kameya et al.: ApJ, **404**, 643 (1993)
9. D. Ward-Thompson: Ap&SS, **239**, 151 (1996)
10. D.Ward-Thompson, P. F. Scott, R. E. Hills, P. André: MNRAS, **268**, 276 (1994)
11. H. Zinnecker, M. J. McCaughrean, J. T. Rayner: 'HH 212: The Most Beautiful Protostellar Jet Known to Date'. In: *Disks and Outflows Around Young Stars, Proceedings of a Conference Held at Heidelberg, Germany 6 - 9 September 1994*, ed. by S. V. W. Beckwith, J. Staude, A. Quetz, A. Natta, (Springer-Verlag Berlin Heidelberg New York 1996)
    Also: Lecture Notes in Physics **465**, 236 (1996)
12. H. Zinnecker, M. J. McCaughrean, J. T. Rayner: Nature, **394**, 862 (1998)

5 Mid-Infrared Imaging of Deeply Embedded Protostellar Sources in the Orion A Cloud

5.1 Abstract

Our goals for this Keck observing run are:

1. Detection and location of heavily embedded sources: mid-infrared imaging is necessary to detect and locate sources which are hidden from view even at near-infrared wavelengths. The use of H_2 jets as pointers to these sources will allow us to construct a well-defined sample of Class I sources.
2. Imaging of deeply embedded multiple sources: most young stars are found in multiple systems (*e.g.* [3]). The high angular resolution of LWS at Keck will allow the examination of multiple systems at their earliest stages. Furthermore, the influence of multiplicity on the formation and evolution of jets and outflows can be studied.
3. 11.7 μm and 18 μm photometry of protostellar jet sources: this will enable us to compile spectral energy distributions of the outflow sources, to put constraints on their evolutionary status, and to detect and to characterize the expected warm, inner protostellar disks.

5.2 Scientific Justification

In our standard picture of star formation, the collapse of a molecular cloud core leads to the formation of a protostellar object, which is initially invisible at optical and near-infrared wavelengths (Class I source). Later on, the obscuring material is cleared away, revealing the young star. It is well known now that outflow activities usually accompany the protostellar evolution even at the earliest stages (*e.g.* [2]). Bipolar molecular outflows are common in the deeply embedded protostellar stage. In fact, these outflows often are the first observable signs of star formation. A number of jets, visible in the near infrared at 2.12 μm (1 - 0 S(1) line of molecular hydrogen), have been found to have their driving source still heavily embedded, *e.g.* the highly symmetric HH 212 jet [8]. The jets are thus an efficient tracer of young (Class I) sources. Many stars at somewhat later evolutionary stages, *e.g.* T Tauri stars, are associated with optically visible Herbig Haro jets. The origin of the outflows as well as the acceleration and collimation mechanisms are still very unclear, but inner accretion disks at a few to a few tens of AU seem to play a crucial role (*jet-disk-connection*, *e.g.* [1]). Thus it is of great importance to identify the sources driving the outflow, to learn as much as possible about their nature, and to relate their properties to the properties of their outflows.

So far, most searches for outflow sources have been rather unsystematic, resulting in inhomogeneous samples of such objects. To get an *unbiased* sample of young stellar objects and associated outflows, we have surveyed a large region in the Orion A molecular cloud (a region of low-mass star formation also known as L 1641) for molecular hydrogen jets. In this region, which covers about one square degree, Tatematsu et al. [7] found 125 dense cloud cores, which are thought to be possible sites of ongoing or future star formation. Our plan is: survey the whole region for protostellar H_2 jets; characterize the jets; relate H_2 jets to molecular outflows and Herbig Haro objects; find and characterize the sources driving the jets (spectral energy distribution $\Rightarrow$ evolutionary stage; properties of protostar/protostellar disk); relate the properties of the cloud cores, the outflows, and the sources driving the outflow. The final goals of this work are to obtain a well-defined sample of protostellar sources in Orion A, and to study the role of jets and outflows in star formation for the first time with a statistical approach.

Our H_2 survey was conducted with the NIR camera OMEGAprime at the 3.5 m telescope, Calar Alto, Spain, using imaging in a narrow band centered on the 2.12 μm transition of H_2, and in K band. Preliminary results from a part of this survey have been published by Stanke et al. [6]. The survey, which is now complete, has lead to the detection of about 40 jet-like H_2 emission features. An additional near-infrared survey (J and H band) is planned for this fall/winter in Spain. In addition, we have recently started to map the whole one square degree region at 1.3 mm with the IRAM 30 m mm-telescope. Identification of the sources driving the jets depends on their evolutionary stage. Rather evolved sources can be identified in the near-infrared images, whereas the very youngest sources may be found only at mm wavelengths. Mid-infrared observations are crucial for detecting jet sources seen neither at near-IR nor mm wavelengths, for constraining the evolutionary stage of mm-sources not detected at near-IR wavelengths (discrimination of class 0 and class I sources), for determining the position of outflow sources not detected in the near-IR, and for providing information at high spatial resolution (multiplicity, presence of small clusters). Our H_2 survey allows us to obtain a good estimate of the source positions (less than 5 arcsec) from the distribution of the H_2 emission (*e.g.* lobes of symmetric jets pointing to a central source, similar to the case of HH 212). This allows targeted observation of these positions, rather than imaging of the full survey area, which would take prohibitively long. During an observing run with the 3.6 m telescope at La Silla (10 μm imaging of the sources of well known protostellar outflows) we also observed (by chance) one source of an outflow (L1641N) which is located in the region of the H_2 survey. In this case, 10 μm imaging already turned out to be a useful tool to locate the outflow source (see [6]).

5.3 Why Do We Need Mid-Infrared Observations at Keck?

- Mid-infrared emission is a tracer of warm dust ($\sim$150 to 300 K; photospheric emission is negligible), which is expected to be found in a disk-like structure around young stellar objects at a distance of a few AU from the central (low mass) object. Therefore mid-infrared emission indicates the presence of inner disks and provides a tool to test the disk-jet-connection.

In addition, near-infrared observations will be used to study the contribution of hot dust and the protostar itself, and millimeter continuum observations will reveal the presence of cold dust.

- Mid-infrared emission penetrates the dusty cloud cores, enabling us to detect heavily embedded objects not visible at optical or even near-infrared wavelengths.
- Keck provides a much higher spatial resolution than space-based observatories (IRAS, ISO, SIRTF). Observations from these missions have poor positional accuracy, which make source identifications doubtful; their poor resolution makes them useless for finding binary sources or small clusters.
- Keck provides much better sensitivity than other ground-based observatories. The good mid-infrared characteristics of the Mauna Kea site allow observations at 18 μm, which is not possible with reasonable efficiency at poorer sites (*e.g.* La Silla). Furthermore, the better spatial resolution helps to resolve clusters or binaries, which is also necessary for the construction of a proper spectral energy distribution.
- Source positions can be measured to an accuracy of better than one arcsecond; thus the array helps to locate sources whose positions are not precisely known (*e.g.* sources surrounded by strong extended reflection nebulosity).
- The spectroscopic capabilities of LWS allow us to determine the strength of silicate features, which are correlated with extinction. This is a particularly interesting measurement for multiple sources, in which differences in evolutionary stage can be revealed in this way.

We propose to use LWS to observe a sample of (candidate) jet sources. Since the final goal of our work (see above) is a statistical study of young stellar objects and their associated outflows, we propose to observe a fairly large sample of $\sim$40 jet sources. The jets/jet sources have been identified during our survey for H_2 jets in the Orion A cloud. In a few cases, IRAS sources with fluxes of a few hundred mJy at 12 μm are found at positions consistent with our jet sources. Since the IRAS fluxes are in many cases contaminated by extended emission, and may reflect the cumulative emission of small clusters, we aim at a substantially better sensitivity, 5 σ detection limits of 10 mJy at 11.7 μm and 30 mJy at 18 μm. The use of these two wavelengths will give us a good handle on the spectral energy distribution, which is necessary to determine the evolutionary status of the sources.

5.4 Technical Remarks

We plan to observe a sample of about 40 sources. We will use two filters with central wavelengths of 11.7 μm and 18 μm. With the best available measurements of the LWS sensitivity (conservatively 40 mJy in 1 s on source at 11.7 μm, 120 mJy in 1 s at 18 μm, see LWS web page), we need 10 minutes

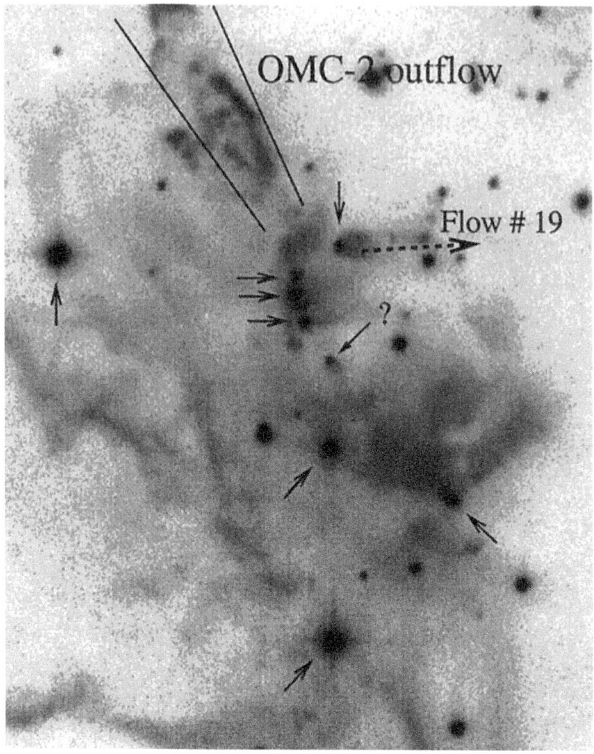

Fig. 14. Near-infrared (2.12 μm) image of the OMC2-cluster (from the Orion A cloud jet survey of Stanke et al. 2000, carried out at the Calar Alto 3.5 m telescope). This image is to be compared with the following mid-infrared image of the same cluster (see Fig. 15) on the same scale.

per source and filter (including a factor of 2 for chopping) to obtain a 5 σ detection limit of 10 mJy at 11.7 μm, and 30 mJy at 18 μm. Allowing sufficient time for target acquisition and setup, we will need about 17 hours to observe our sample. In addition, we intend to perform pilot spectroscopic observations (low-resolution, N band) of a few of the brightest sources. Orion A can be observed for about 7 hours per night at Keck Observatory, which means that we will need three nights for our program. The time before and/or after Orion is visible will be used for observation of photometric standard stars.

5.5 Results

Guided by the geometry and morphology of embedded jets previously found in our large-scale unbiased H_2 jet survey in the Orion A cloud carried out with a wide field near-infrared camera at the Calar Alto 3.5 m telescope, follow-up observations with a mid-infrared camera at the 10 m Keck telescope on

Mauna allowed us to detect several faint 10 μm point sources, likely to be the obscured star-disk energy sources for the jets. A similar project conducted with camera at the ESO 3.6 m telescope revealed only the brightest 10 μm jet-driving protostars (see Fig. 15). The 10 μm emission most likely results from warm (300 K) dust in the inner circumstellar disk from which the obscured young stellar objects are accreting.

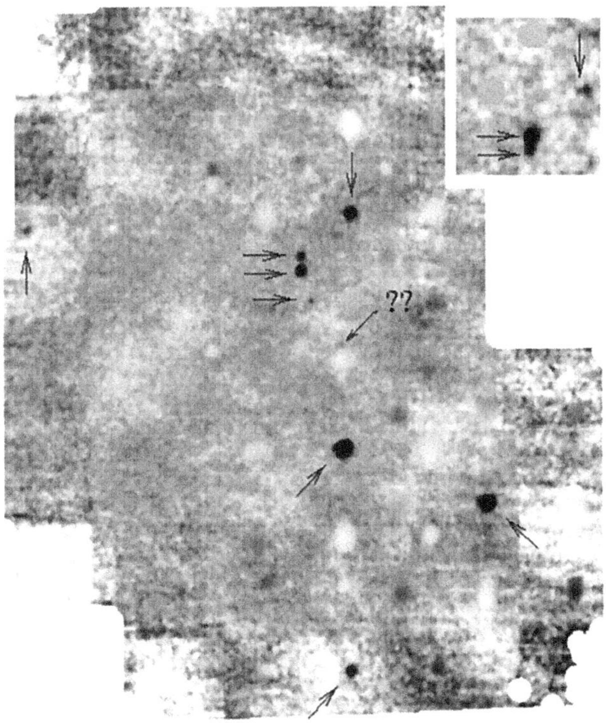

Fig. 15. Mid-IR (11.8 μm) image of the OMC-2 cluster ($\sim$ 2.4 arcmin $\times$ 3.0 arcmin) obtained using TIMMI2 at the ESO 3.6 m telescope. A variety of sources are seen in the 11.8 μm mosaic, which in this case all have a 2.1 μm counterpart (see Fig. 14). Additional 18 μm images (see insert, top right) demonstrate that the northern component of the close pair actually becomes the brighter source at this wavelength; it is reasonable to assume that the northern, brighter source also is the dominant source of luminosity at mid- to far-IR wavelengths. *Presumably this is the source driving the OMC-2 molecular outflow*, which is visible in the 2.12 μm image extending towards the north-east (see Fig. 14). Another 11.8 μm source is seen to the north-west of the close pair; this source drives another outflow (marked as Flow # 19 in Fig. 14). (This figure was kindly prepared by Th. Stanke.)

It is clear that only high spatial resolution mid-infrared imaging can help to constrain the spectral energy distributions of sources like these at mid-infrared wavelengths: the resolution of IRAS is far too poor to resolve a little cluster like this.

References

1. A. Königl: 'Theory of Outflows', Review. In: *Disks and outflows around young stars, Proceedings of a Conference Held at Heidelberg, Germany 6 - 9 September 1994*, ed. by S. Beckwith, J. Staude, A. Quetz, A. Natta (Springer-Verlag Berlin Heidelberg New York 1996)
Also: Lecture Notes in Physics **465**, 282 (1996)
2. C. J. Lada: ARAA **23**, 267 (1985)
3. R. D. Mathieu: ARAA **32**, 465 (1994)
4. A. Quirrenbach: 'Operating the VLTI in the thermal infrared'. In: *Science with the VLT Interferometer, Proceedings of the ESO workshop, held at Garching, Germany, 18-21 June 1996*, (Springer-Verlag, Berlin, New York 1997) p. 385
5. A. Quirrenbach, H. Zinnecker: 'Adaptive optics observations of molecular hydrogen towards T Tauri'. In: *Adaptive Optics and Applications*, SPIE Vol. **3126**, 249 (1997); see also Msngr **87**, 36 (1997)
6. T. Stanke, M. J. McCaughrean, H. Zinnecker: A&A **332**, 307 (1998)
7. K. Tatematsu: ApJ **404**, 643 (1993)
8. H. Zinnecker, M. J. McCaughrean, J. T. Rayner: Nature **394**, 862 (1998)

6 A Study of Faint Disks Around X-ray Selected Young Low-Mass Stars in an OB Association

6.1 Abstract

We propose an ISOCAM/ISOPHOT search at 6.7/15 μm and 60/100 μm for faint circumstellar disks around very young (1 - 2 Myr) stars of low mass in the most nearby OB association (Sco-Cen or Orion depending on the launch window). All these stars are weak-line, X-ray (EINSTEIN, ROSAT) selected T Tauri stars (WTTS). It has been suggested that such stars are truly naked (diskless). The sensitivity of ISO is such that we can probe the presence of disk dust masses up to a factor of 100 lower than ground-based observations (*i.e.* of order earth masses). By concentrating on the youngest WTTS, we will find out whether WTTS are genuinely naked or whether WTTS only lack the innermost disk and, correspondingly, the signatures of accretion (*e.g.* large Hα equivalent widths). Therefore the proposed observations go a long way to understand the difference between weak-line and classical T Tauri stars. Indeed we expect to find faint dusty disks and will investigate their properties (dust mass, dust temperature) in a well-defined nearly coeval sample (each object is placed in the HR-diagram). The 7/15 μm data will probe the presence of warm inner disks at radii of the order 1 AU, important for the formation of terrestrial planets. The 60/100 μm data bear on the presence of cold outer disks with a scalelength of the order of 100 AU, important for the formation of very low-mass binary companions. Models suggest the disks are likely to be optically thin at each of the 4 wavelengths, in which case the dust mass of both the inner and outer disks can be determined. This will respresent an enormous quantitative advance in our understanding of young disks (disk dissipation timescale as a function of stellar mass), in an environment where most of the field stars are believed to form.

6.2 Description of the Proposed Programme

Background and Open Problems. Most young stars, including low-mass stars, seem to form in OB associations. Already in 1978 Miller and Scalo [8] pointed out that if the IMF of high-mass stars defining an OB association is extrapolated to low-mass stars (then not yet detected), then OB associations would be the primary birth place of the low-mass stars in the field. While young low-mass stars were long known in T associations (like Taurus), the existence of young low-mass stars in OB associations (like Orion or Sco-Cen) has just been verified through recent all sky X-ray observations (in particular ROSAT). X-rays are a powerful pointer to identify young low-mass stars, something that became clear a decade or so ago after the first imaging X-ray data obtained with the EINSTEIN Observatory were analysed (*e.g.* [11]). Because young stars are rapid rotators and are also convective, they should have magnetic dynamos and bright coronae, giving rise to strong X-ray emission. Thus X-ray emission is an efficient method to discriminate between old Main Sequence and young pre-Main Sequence stars of low-mass, and hence to discover candidate young low-mass stars. It turns out that many of these candidates are indeed pre-Main Sequence objects, as judged from their placement in the HR-diagram, with only weak Hα emission lines (therefore the name weak-line T Tauri stars or WTTS). Therefore they did not show up in objective prism emission line surveys for young low-mass stars (those stars are the so-called classical T Tauri stars or CTTS, which in addition also have NIR and UV excess emission). Now one can ask why WTTS are so inconspicuous in Hα or in NIR and UV. The current concensus is that this is because, contrary to the classical T Tauri stars, they have no or only weak circumstellar disks. The question then is why are there two populations of T Tauri stars, the classical ones and those that only were discovered through their X-ray emission? One answer may be that the strong magnetic fields of the X-ray selected PMS stars can destroy the inner part of their disks (perhaps early on). Another answer may be that the X-ray selected stars are much older than the classical TTS and that therefore their disks have evolved and dissipated (*e.g.* by accretion earlier on). A third possibility is that, although these objects may not have inner disks, they still have outer disks (*e.g.* the dust in the inner disk has been depleted by the Poynting-Robertson drag, by the action of a close binary system, or by planet formation). However, the best ground-based 1.3 mm and 0.8 mm dust continuum observations have failed to detect these outer dust disks (*e.g.* [2]; [9]). This has led some people to call these objects "naked" T Tauri stars (Walter). But are WTTS really naked? One might expect that at least in the youngest systems these outer dust disks should be present, unless these systems somehow form without any disks at all - a revolutionary idea, unlikely but worth testing.

As of now, we do not understand the physical connection between the classical and the weak-line T Tauri stars. This is all the more true since CTTS and WTTS often occupy the same part in the HR-diagram, *i.e.* the same variety of ages and masses. This suggests that the difference is not purely an evolutionary one, but maybe genetic. It is of great astronomical significance to understand the origin of this difference, especially since the WTTS seem to outnumber the CTTS, thus the WTTS play the bigger role for the Initial Mass Function. It is conceivable that the difference between WTTS and CTTS has something to do with binary companions at small separations (*e.g.* [4]), although this is controversial (*e.g.* [6]). Binary companions at small/large separations would effect the inner/outer disks considerably, reducing the dust mass [2].

Sensitive ISO observations can probe the presence, the amount, and the temperature of the dust in the inner/outer disk of X-ray selected WTTS. Model calculations are available to interpret any dust emission spectra that will be observed (including various types of dust grain composition).

By focussing on the youngest WTTS (1 - 2 Myr) we will find if the WTTS are born naked (ie devoid of dusty circumstellar material) or whether so far we just lacked the sensitivity to detect their faint dusty disk emission.

Need for ISO Data. The need for ISO data is due to the faintness of the dusty disks of WTTS. ISO observations both in the short wavelength regime (6.7/15 μm) and in the long wavelength regims (60/100 μm) are about 100 times more sensitive to the total mass of warm/cold dust (assuming optically thin emission) than current ground-based instruments (*e.g.* TIMMI at ESO for 10 μm and the mm wave telescopes detecting cold dust emission, such as CSO, JCMT, SEST, and even IRAM). Furthermore, due to being a cooled infrared telescope, ISO is also typically a 100 times more sensitive than IRAS. For the study of faint disks ISO represents a giant step forward.

Technical Feasibility. We have selected 40 targets in the Sco-Cen association (Upper Sco subgroup): 24 from Walter *et al.* [12] and 16 from Kunkel [5]. In both cases, we are dealing with X-ray selected targets, either following EINSTEIN pointings (Walter *et al.*) or ROSAT pointings (Kunkel). The corresponding spectroscopic follow-up observations of the X-ray sources provided not only identification as weak-line T Tauri stars but also spectral types (effective temperatures) and luminosities, so that each young stellar object in our target list has been positioned in the HR-diagram. Using pre-Main Sequence evolutionary tracks, such as those by d'Antona and Mazzitelli [1], each star in our target list has a rather well-defined mass and age. Since we are dealing with weak-line T Tauri stars, both masses and ages are not corrupted by an excess luminosity due to active accretion onto the stellar surface. In fact, the present proposal is chasing "passive accretion disks". Such passive disks are faint, and indeed they have not been detected in 1.3 mm dust continuum observations ([*e.g.* [9]). The sensitivity of ground-based searches for dust disks is 0.1 Jy at 1.3 mm (with 15 m class telescopes like JCMT or SEST; somewhat better with IRAM 30 m) which implies of order 10 Jy at 100 μm for optically thick emission. Inspection of the IRAS PSC reveals that no such flux density has been seen; indeed the sources are also not detected in the IRAS FSC at an upper limit of around 1 Jy. Therefore we are challenged to search for faint dust disks at the 0.1 Jy level. This necessitates the ISOPHOT_C100 observations. The ISOCAM observations near 10 μm aim at detecting the photospheres of naked M and K pre-Main Sequence stars in the Sco-Cen OB associations. (Any emission in excess of the photospheric flux would be due to a disk.) K and M are the typical spectral types for the young stellar counterparts of the X-ray sources in this nearby (160 pc) region. [In the more distant Orion region the dominant spectral type would be G, but the requirement to detect a G2V star photosphere in Orion is about the same as it is for the detection of an MOV photosphere in Sco-Cen]. The sensitivity needed is in the range 1 - 5 mJy (*cf.* [10]), 1 mJy corresponding to an M4 star at a distance of 160 pc. 1 mJy can be achieved with the ISOCAM LW2 and LW3 in 20 readouts of 10 sec each (with a pixel FOV = 1.5 arcsec).

Related ISO Observing Programmes. There are already several disk projects in the ISO central programme, notably those by Beckwith *et al.* and Ray *et al.*; Beckwith *et al.* aim at studying the evolution of disks as a function of time (*e.g.* searching for fainter and fainter disk in progressively older clusters). This is not in conflict with us, since we perform something complementary: searching for disks in a coeval population of (very) young stars in an effort to understand disk origin and dissipation as a function of mass. As for comparison with the Ray *et al.* project which does include WTTS targets, we point out that our sample is much more homogeneously selected; we know the mass and age of every single object in our sample from the HR-diagram, and we emphasize the particular aspect of choosing extremely young weak-line T Tauri stars (for reasons that are explained above). This should enable us to reach more powerful conclusions than Ray *et al.*.

6.3 Results

We report the discovery of weak thermal infrared excess in several weak-line T Tauri stars in the Scorpius-Centaurus OB association, obtained with ISOCAM aboard the ISO satellite. This excess, measured at 6.7 and 14.3 μm, suggests the presence of faint remnant disks, with sizes of the order 1 AU. These young stellar objects are about 5 Myr old and were previously believed to be "naked", *i.e.* lacking a circumstellar disk. Thus, planet formation may be ongoing around some Sco-Cen weak-line T Tauri stars, whose masses are around a solar mass. (See: W. Brandner, H. Zinnecker, A. Moneti et al.: 'Timescales of Disk Evolution and Planet Formation: HST, Adaptive Optics, and ISO Observations of Weak-Line and Post-T Tauri Stars', AJ **120**, 950 (2000))

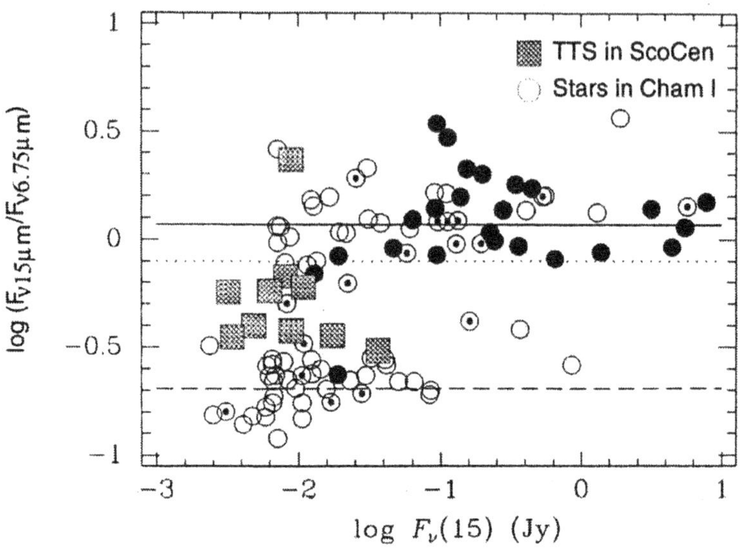

Fig. 16. Color-magnitude diagram based on ISOCAM observations of young stars in Chamaeleon (circles, Nordh et al. 1996) and Scorpius-Centaurus (squares, this work). Black filled circles indicate previously known YSOs and CTTSs, circles with a central dot previously known WTTSs, and open circles new sources detected with ISO by Nordh et al. (1996). The dashed line indicates the location of pure stellar photospheres, the dotted and solid line the location of flat and flared circumstellar disks, respectively, as predicted by the models from Kenyon & Hartmann (1995). WTTSs and PTTSs in Sco-Cen show a spectral index intermediate between main-sequence stars and CTTSs and WTTSs in Chamaeleon. (Courtesy A. Moneti)

References

1. F. D'Antona, I. Mazzitelli: ApJS **90**, 467 (1994)
2. S.V.W. Beckwith, A.I. Sargent, R.S. Chini, R. Güsten: AJ **99**, 924 (1990)
3. E.J. de Geus: AA **262**, 258 (1992)
4. A.M. Ghez, G. Neugebauer, K. Matthews: AJ **106**, 2005 (1993)
5. W.E. Kunkel: Untersuchungen massearmer Sterne in der Scorpius-Centaurus Assoziation mit ROSAT. PhD Thesis, Universität Würzburg, Würzburg (1996)
6. Ch. Leinert, H. Zinnecker, N. Weitzel et al.: AA **278**, 129 (1993)
7. M.R. Meyer, B.A. Wilking, H. Zinnecker: AJ **105**, 619 (1993)
8. G.E. Miller, J.M. Scalo: PASP **90**, 506 (1978)

9. S. L. Skinner, A. Brown, F. M. Walter: AJ **102**, 1742 (1991)
10. M. F. Skrutskie, D. Dutkevitch, S. E. Strom et al.: AJ **99**, 1187 (1990)
11. F. M. Walter, L. V. Kuhi: ApJ **284**, 194 (1984)
12. F. M. Walter, F. J. Vrba, R. D. Mathieu, A. Brown, P. C. Myers: AJ **107**, 692 (1994)

III. Young Binaries

7 Statistics of Young Binary Stars: Implications for (Binary) Star Formation*

7.1 Young Binary Stars: Why Care?

It is well known that most main sequence stars are members of binary and multiple systems (*e.g.* [2]; [19]). Recent survey work in nearby star forming regions (such as Taurus and Ophiuchus) has demonstrated that this is also the case for young low-mass pre-main-sequence stars [41]; [35]; [25]; [14]; [38]. Therefore, it is fair to say that binary star formation is the rule in star formation, while single star formation (*cf.* [36]) seems to be the exception. Statistical studies of pre-main-sequence binary stars are privileged over those of Main Sequence stars in that the former generally carry a closer memory of star formation and of binary formation mechanisms in particular. Also the memory of environmental influences on binary star formation (differences among young binaries born in T *vs.* OB associations or young star clusters) is not yet lost. We can also turn the question around and ask if there is something about Main Sequence binaries that we can use to learn about binary star formation without resorting to young pre-main-sequence objects. Indeed, the distribution of mass ratios might be such a property, provided mass ratios are frozen in since stellar birth. This is because on the Main Sequence there is a well-calibrated luminosity mass relation, while this relation is time-dependent for contracting pre-main-sequence objects; furthermore, accretion luminosity can contaminate the stellar luminosity in the case of actively accreting T Tauri stars, messing up any mass-luminosity relation.

Figure 17 shows an example (Sz 24) of a T Tauri star binary system resolved by direct CCD imaging at the 3.5 m ESO New Technology Telescope at La Silla. The component separation is 0.69 arcsec. This is just wide enough to secure optical spectra of the individual components enabling us to place the two components separately on the H-R diagram. A number of other similar binary objects could also be placed on the H-R diagram Fig. 18. One can easily see that if young binaries are not recognized as such, the masses and ages of these systems would be substantially misjudged (*cf.* [37]). This then shows that unresolved young binaries can cause errors in the determination of the Initial Stellar Mass Function and also in the reconstruction of the star formation history.

7.2 Observational Methods

Figure 19 shows the frequency distribution as a function of semimajor axis for solar-type main-sequence binaries (adapted from [12]). If we assume the

* Paper reprinted from Proc. Conference "Visual Double Stars: Formation, Dynamics, and Evolutionary Tracks", A. Docobo et al. (eds.), p. 115 (1997)

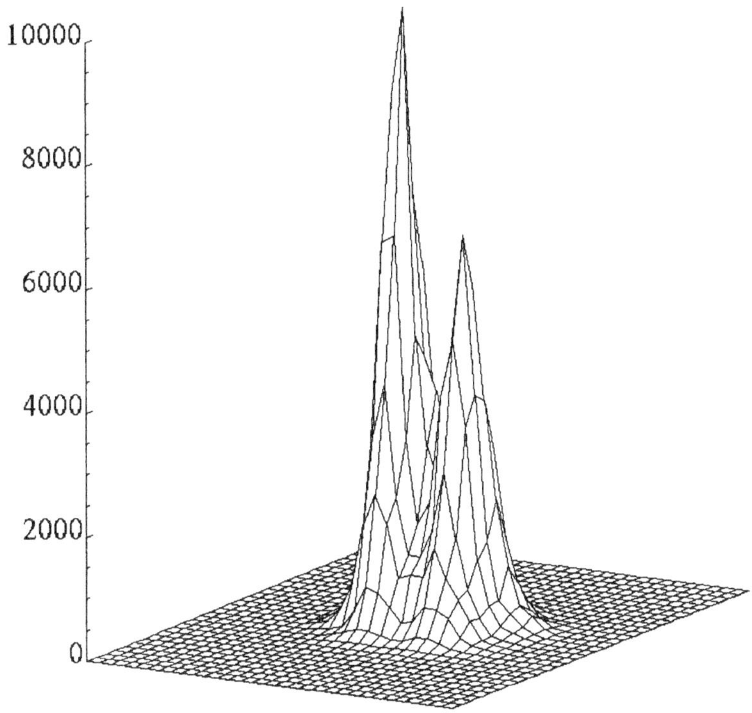

Fig. 17. The young visual binary star Sz 24, also known as VW Cha, resolved into its components (projected separation 0.69 arcsec).

same distribution to be valid for T Tauri stars at a distance of 150 pc, we can estimate which percentage of binaries remains unresolved. Using conventional observing techniques (*i.e.* direct imaging) under typical seeing conditions (FWHM 1 arcsec) we detect only 30 % of all binaries.

In order to achieve a higher spatial resolution more sophisticated observing techniques such as speckle imaging or lunar occultation need to be employed. In the following we give a brief summary of the different techniques used to search for PMS binaries.

1 arcsec resolution: *direct imaging* in the optical and near infrared offers seeing limited resolution.

0.1 arcsec resolution: *speckle imaging & adaptive optics* both compensate for the wavefront distorsions induced by earth's atmosphere and produce diffraction limited images (*e.g.* $\lambda/\mathrm{D} \approx 0.1$ arcsec for $\lambda = 2.2\,\mu$m and $\mathrm{D} = 4$ m telescopes).

0.01 arcsec resolution: *lunar occultation & HST-Fine Guidance Sensors* provide even higher resolution independent of the seeing, but

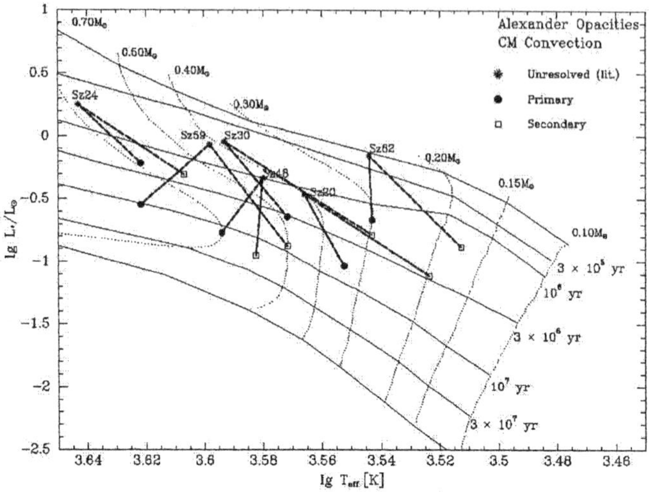

Fig. 18. A sample of young binaries in the H-R diagram before and after resolving them into their individual components. It is seen that unresolved binaries lead to an underestimate of the age of a T Tauri star population [9].

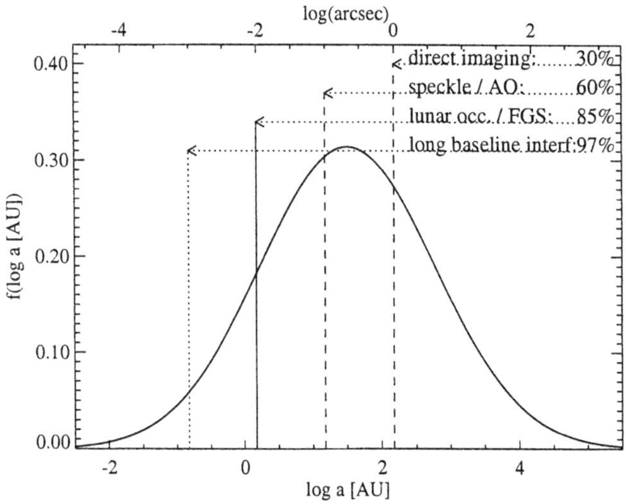

Fig. 19. Distribution of semimajor axes for nearby solar-type main-sequence binaries according to Duquennoy & Mayor [12]. The resolution limits of various detection techniques are indicated for an assumed distance of 150 pc (*i.e.* the distance to the nearest star forming regions).

dependent on the diameter of the telescope and system throughput (effectively the total light collecting power).

0.001 arcsec resolution: *long baseline interferometry* is currently only applicable to bright sources. The new interferometers now under construction (Keck, CHARA, VLTI, LBT) will be able to resolve PMS binaries with orbital periods of a year or even less (a separation of 0.007 arcsec corresponds to 1 AU at a distance of 150 pc).

7.3 Results

- The *binary frequency* amoung pre-main-sequence stars is at least as high as among main-sequence stars. Individual low-mass star forming regions (*i.e.* Taurus-Auriga) show a clear (factor of 2) overabundance of pre-main-sequence binaries with separations $\geq$ 15 AU in comparison to the main-sequence [25]; [14]; [16]. Other regions (*i.e.* Ophiuchus, Trapezium cluster) appear to have a similar binary frequency as on the main sequence [38]; [34]. For binaries with separations $\geq$ 150 AU the overabundance is mild or non-existent within the statistical uncertainties [35]; [29]. However, there are variations in the binary frequency within the boundaries of certain associations, such as the Sco-Cen OB association; here localized areas exist where the PMS binary frequency is enhanced or suppressed with respect to the main sequence [8].
- The histogram of infrared (1μm) *brightness ratios* I_2/I_1 (Fig. 20a) of the components of wide visual PMS binaries ($\geq$ 150 AU) exhibits a small gradual increase towards smaller ratios, the median being around 0.3 [35]; [8]. The corresponding histogram of PMS speckle binaries (now measured at 2 μm) looks flat for ratios I_2/I_1 in the range 0.5 - 1.0 [25]. The distribution seems to rise slightly (Fig. 20b) for smaller ratios, but this may be an artifact which goes away when a coarser binning is used. At any rate, there is a selection effect against the detection of the smallest brightness ratios (first bin in Fig. 20b) at the smallest separations (near the diffraction limit). Therefore the jury is still out regarding the frequency distribution of brightness ratios for PMS stars with faint companions (cf. [14]; [16]).
- The *surface density of companions* for the 40 young Taurus speckle binaries rises with the 2.4 power towards smaller projected separations [25] (see also [24] who finds a slope of 2.15 when combining the data sets of Leinert *et al.* [25]; [14]). Thus the corresponding *semi-major axis distribution* of nearby PMS binaries roughly follows a 1/a distribution in the range of separations 0.1 - 12 arcsec, *i.e.* 15 - 1800 AU. Similarly, the systematic CCD survey among visual T Tauri stars in southern star forming regions produced a sample of 42 visual binaries and 1 triple system, whose distribution of separations also follows a 1/a law between 150 - 1800 AU

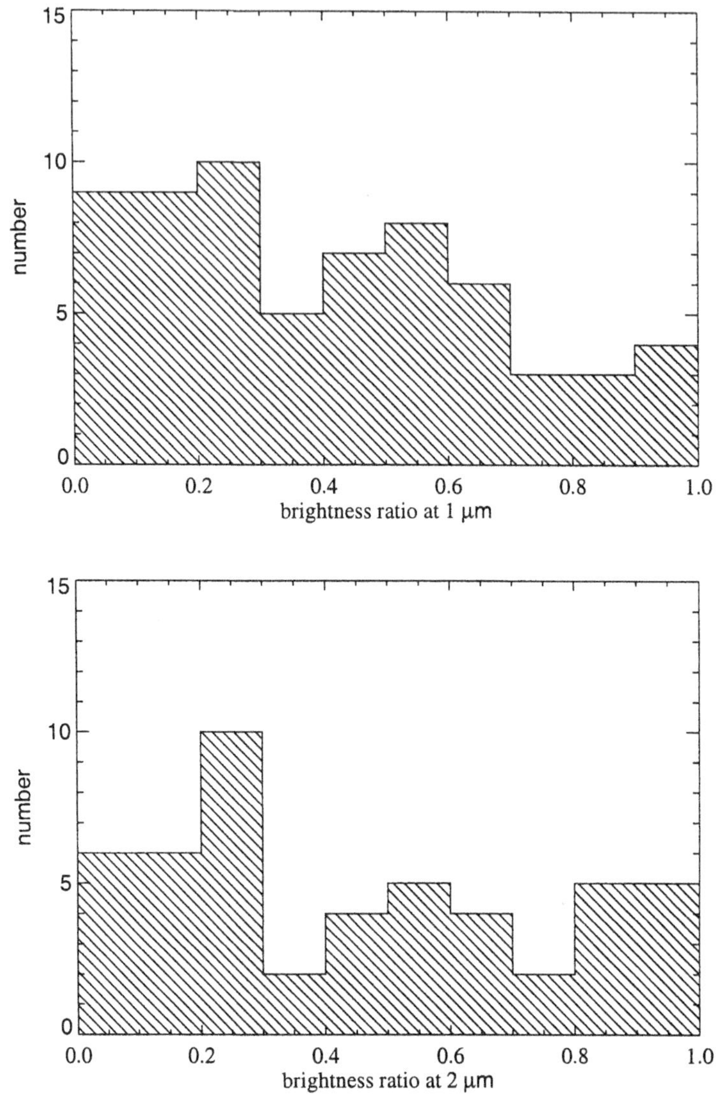

Fig. 20. Top: Brightness ratio I_2/I_1 of the components of visual pre-main-sequence binaries at 1 μm [35]. Bottom: Brightness ratio I_2/I_1 of the components of speckle pre-main-sequence binaries at 2 μm [25].

[35]. This is essentially Öpik's law, originally found and confirmed for main sequence binaries (see Allen, Poveda & Herrera, *ibid.* p. 133).

- The components of almost all close visual pre-main-sequence binaries appear to be coeval to within 1 Myr or less [9], see Fig. 18). Previous work on this topic [18] which concluded that only 2/3 of the chosen PMS binary sample had *coeval components* used much wider visual binaries some of which may be optical pairs.
- There is a subclass of young binaries in which some optically visible T Tauri stars have an *infrared companion* (*i.e.* a companion which is invisible or just barely visible in the optical). Examples include the T Tau system itself, Haro 6 - 10, UY Aur, and a few others such as VV CrA (Fig. 21). For an early review see [42], while the most recent paper is by Koresko, Herbst, & Leinert [22]. In particular, the latter authors noted that UY Aur changed its state from visible to infrared companion in the last 50 years or so. Therefore, based on the observed variability, they propose that infrared companions are objects which undergo episodic accretion events, similar to FU Orionis outbursts. In this context, it may also be relevant that in classical T Tauri binaries it is the infrared bright component that carries most of the infrared color excess [31].
- Double jets emanating from embedded binaries are not aligned (*e.g.* HH 111/HH 121, [17]; see Fig. 22). Other examples can be found in [20] and in [21]. This strongly suggests that the planes of *circumstellar disks* around the individual components of a pre-main-sequence binary are not coplanar [40], contrary to the simplest rotational fragmentation schemes.
- Recently, Leinert, Haas, & Richichi [26] extended PMS binary statistics to some 30 *Herbig Ae/Be stars.* Although their sample is inhomogeneous and incomplete (*i.e.* objects at various distances are included), their initial survey indicates that Ae/Be stars have similar binary frequencies as T Tauri stars. As for the binary frequency of (young) massive *O-type stars*, it is being investigated by the CHARA group and some results of their survey of 239 objects ($V < 8$) have been announced [28]. It is clear that again the binary frequency is "high".
- Latest news: 2 μm adaptive optics observations of the *Trapezium stars* in the Orion Nebula carried out at the ESO 3.6 m telescope with ADONIS have shown θ^1A to be a 0.2 arcsec binary (a result first found by Petr *et al.* 1996 with speckle techniques). Moreover, θ^1B turned out to be a 0.7 arcsec binary, with the secondary being a (very red) 0.2 arcsec binary itself. Thus θ^1B is a spatially resolved triple system. Finally, θ^2A is a sub-arcsec binary (0.3 arcsec), too [30].

7.4 Implications

1. Binaries form prior to the pre-main-sequence stage. This and the coevality of the components argues against random pairing of field stars and in

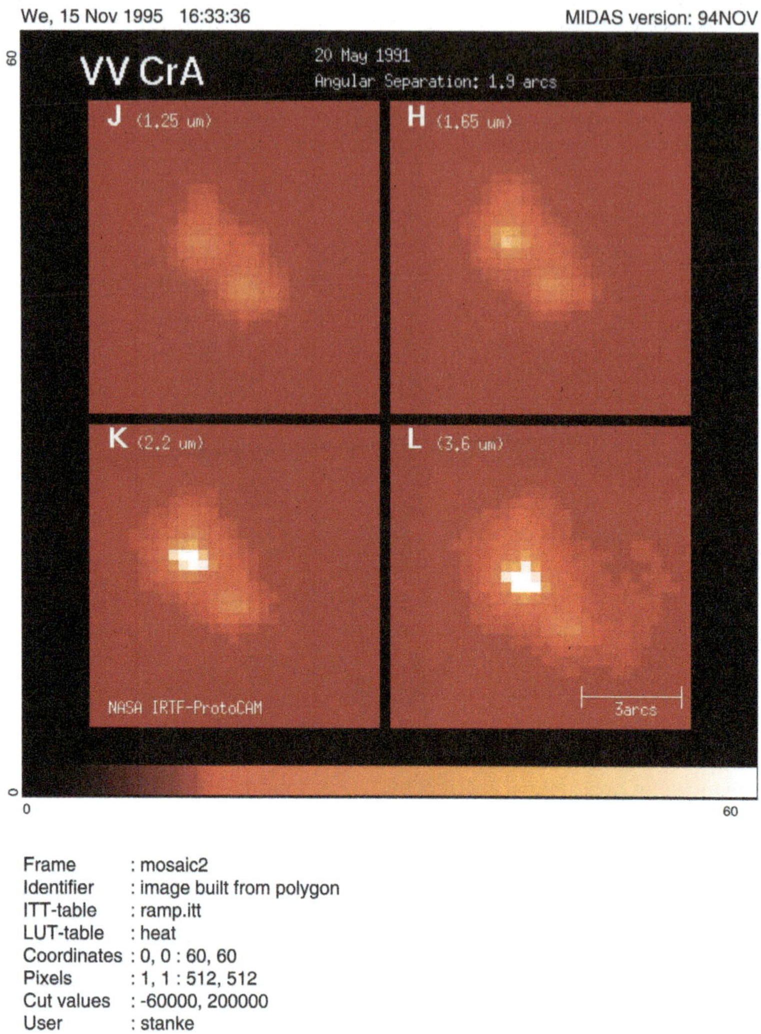

Fig. 21. Infrared array images of the VV CrA young binary system (from Stanke & Zinnecker, unpublished). The infrared companion becomes increasingly brighter at longer infrared wavelengths. At optical wavelengths (not shown here) the infrared companion is barely visible.

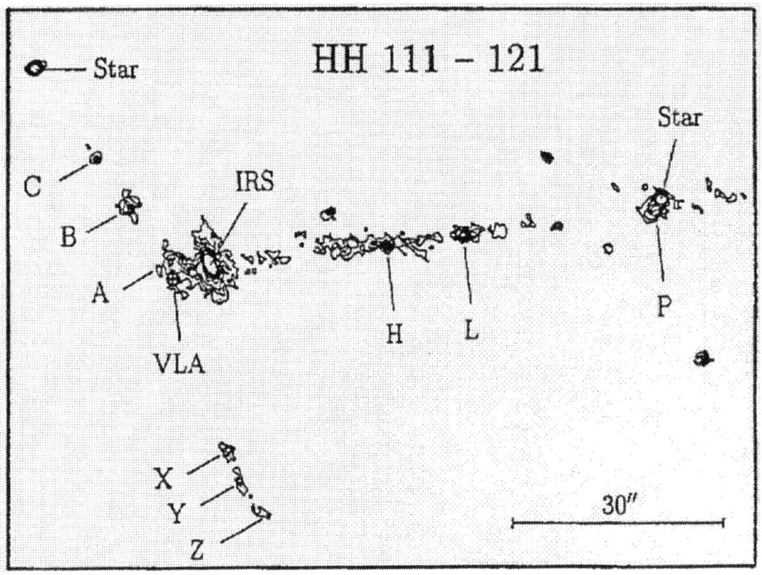

Fig. 22. Jets and Herbig-Haro objects arising from the individual components of pre-main-sequence binaries appear to be misaligned. The 2 μm image above (taken from [17]) shows two series of knots of shock excited molecular hydrogen emission at an angle of $\approx 120°$, emanating from a deeply embedded putative binary system in Orion (IRAS 05491 +0247). The misalignment indicates that circumstellar disks in binary systems may not always be coplanar.

favour of some sort of correlated formation (*i.e.* fragmentation during collapse or independent condensation in small groups).

2. The distribution of mass ratios is consistent with independent pairing from a Miller Scalo IMF (*cf.* [27]) and perhaps suggestive of negative feedback effects due to accretion in protobinary systems (*cf.* [4]).
3. The distribution of semi-major axis of visual PMS binaries, taken at face value, is scale-free (1/a distribution) down to the resolution limit (15 AU). Alternatively the observed distribution is also consistent with a Gaussian in log a, with a peak around 30 AU (*cf.* [12]). It is important to discriminate which of the two shapes is the more correct, as each shape requires a different type of explanation. Several theories for the origin of semi-major axis or period distribution have been suggested depending on the adopted shape. A broad Gaussian in log a can be explained by models combining magnetic braking and ambipolar diffusion in protostellar clouds [32] or by models of a multi-step hierarchical fragmentation [5]; [39]. On the other hand, the scale-free 1/a distribution – valid over a wide range (two orders of magnitude) – may point to orbital decay induced by

gravitational drag (dynamical friction) and related angular momentum transport effects [24].
4. Binary components possess circumstellar disks. This suggests that planet formation can take place in sufficiently wide binary systems, even though the planets may not be stable and may be ejected (free-floating planets).
5. The star formation environment is important for the binary frequency. The binary frequency can differ in different star forming regions. This may allow to trace the main origin of the field star population ("inverse binary population synthesis").
6. The difference between CTTS and WTTS (presence or absence of accretion disks) is not primarily due to binarity.
7. The origin of the infrared companions is still a matter of debate. Bonnell & Bastien [6] and Bate & Bonnell [4] relate the infrared companion to accretion events onto the more massive binary component, while Artymowicz & Lubow [3] and Koresko, Herbst, & Leinert [22] favor the less massive of the two components to be the accretor and hence the infrared companion. While Koresko *et al.* may well be on the right track with their variability theory for infrared companions (see above), differential mass-dependent evolutionary effects as well as differential disk inclination effects may also play a role [42].

7.5 Future Prospects

Up to now, mass estimates for PMS stars are primarily based on the comparison with theoretical PMS evolutionary tracks in an H-R diagram. Different sets of theoretical PMS evolutionary tracks based on different input physics (*e.g.* [11]; [13]; [33]), however, yield ambiguous results. In addition, star-disk interactions (accretion luminosity) and uncertain intrinsic colours for young stars makes it often difficult to place these stars in an H-R diagram at all.

With a typical distance of 150 pc to the most nearby star forming regions and a diffraction limit of 0.1 arcsec at 2.2 μm at a 4 m class telescope, only pre-main-sequence binaries with separations $\geq$15 AU have been resolved so far. Continuous observations over a period of 50 yr or even longer are necessary in order to determine the orbital parameters of these binaries. The new generation of 8 m to 10 m class telescopes and the large optical interferometers now under construction have the power to resolve PMS binaries in nearby star forming regions with semimajor axis of 1 AU and less. This will enable us to determine orbital parameters, mass ratios, and individual masses of pre-main-sequence stars within a timespan of a year. Thus, these observations will allow for a mass calibration of (theoretical) PMS tracks.

In addition, these very high angular resolution observations will allow us to decide whether the semi-major axis distribution is unimodal or bimodal. From this we will ultimately learn whether there is a single binary formation mechanism which can explain the full range of component separations or two mechanisms – one for wide binaries (filament fragmentation?) and one for close binaries (disk fragmentation?).

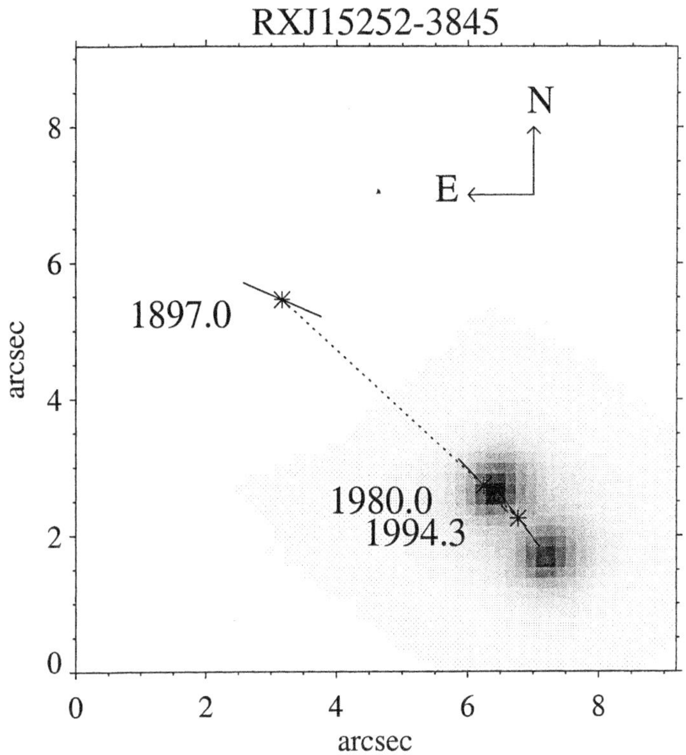

Fig. 23. Shown in gray scale is the binary system at epoch 1994.3 (separation 1.05 arcsec), and its common proper motion in the sky by about 5 arcsec over the last 100 years or so (dashed line). The center of light at 3 epochs is indicated with a star symbol, while the binary separation and position angle is given by the short line. Note that the position angle at epoch 1897.0 was 247 deg, different from the later epochs: hence there is evidence for orbital motion (cf. [8]).

7.6 Discussion

McAlister: Have orbital motions, rather than just relative motions, yet been detected? Isn't there still a chance that some of these are optical?

Zinnecker: This is an important question. For speckle binaries at a distance of 150 pc and components separation of the order of 0.1 arcsec (15 AU), the orbital period is 60 yr, *i.e.* within 5 years significant orbital motions should be detectable. However, for wider visual binaries with separations of order 1 arcsec (150 AU), the orbital periods exceed 1000 yr and orbital motions cannot be detected within a few years. – Yet for some such binaries, *e.g.* the weak-line T Tauri star RXJ 15252 - 3845 (= IDS 15189 - 3823 BC = HD 137059) data with an epoch difference of 100 yr is available and orbital motion can indeed be seen (see Fig. 23)!

Bonnell: I'd like to make two comments. 1) If a binary forms from fragmentation during collapse, there is no reason why the different parts of the cloud can't have unaligned angular momenta vectors such that the resultant binary has unaligned rotation axes. 2) Although orbital motions cannot be shown for individual PMS speckle binaries, the work of Ghez *et al.* [15] shows that statistically it is consistent with orbital motion in that more massive and closer systems move more quickly.

Beust: You said that up to now no young eclipsing binary has been discovered. But I will give a talk tomorrow morning on such a system, namely TY CrA. Secondly, you mentioned that binaries should normally form with aligned rotation axes. This may be true for isolated binary systems, but not for multiple ones. Here again, I will show that concerning the TY CrA system, which is in fact a triple system, we have good reason to think that the rotation axes of the central binaries are not aligned, thanks to the pertubations of the third body.

Abt: There is interesting work being done in N-body simulations relevant to binary formation. Perhaps the most pertinent study was by Aarseth & Hills [1], which involved a range of stellar masses and subclusters to give faster rates. It showed that binaries formed and were disrupted; new pairs were formed and were disrupted, so that the final pairing bore little resemblence to the initial pairing.

Zinnecker: Yes, if most stars in the Galaxy were born in initially bound clusters (consistent with current results from infrared imaging), cluster dynamics may well have a substantial effect on the final binary pairings. There are 3 new papers (ideas) on the effects of clusters (small N, large N, hierarchical or otherwise) on binary star formation to which I would like to refer to [10], [23], and [24].

References

1. S. J. Aarseth, J. G. Hills: A&A **21**, 225 (1972)
2. H. A. Abt: ARAA **21**, 343 (1983)
3. P. Artymowicz, S. H. Lubow: ApJ **421**, 651 (1994)
4. M. R. Bate, I. A. Bonnell: MNRAS **285**, 33 (1997)
5. P. Bodenheimer: ApJ **224**, 488 (1978)
6. I. A. Bonnell, P. Bastien: ApJ**401**, 654 (1992)
7. I. A. Bonnell, M. R. Bate: MNRAS **269**, L45 (1994)

8. W. Brandner, J. M. Alcalá, M. Kunkel, A. Moneti, H. Zinnecker: A&A **307**, 121 (1996)
9. W. Brandner, H. Zinnecker: A&A **321**, 220 (1997)
10. C. J. Clarke: 'The Formation of Binaries in Small N Clusters'. In: *The origins, evolution, and destinies of binary stars in clusters, Astronomical Society of the Pacific Conference Series, Volume 90, An international symposium held at the University of Calgary, 18-23 June 1995*, ed. by E. F. Milone, J.-C. Mermilliod (Astronomical Society of the Pacific, ASP, San Francisco 1996), p. 242
11. F. D'Antona, I. Mazzitelli: ApJS **90**, 467 (1994)
12. A. Duquennoy, M. Mayor: A&A **248**, 485 (1991)
13. M. Forestini: A&A **285**, 473 (1994)
14. A. M. Ghez, G. Neugebauer, K. Matthews: AJ **106**, 2005 (1993)
15. A. M. Ghez et al.: AJ **110**, 753 (1995)
16. A. M. Ghez, D. W. McCarthy, J. L. Patience, T. L. Beck: ApJ **481**, 378 (1997)
17. R. Gredel, B. Reipurth: ApJ **407**, L29 (1993)
18. P. Hartigan, K. M. Strom, S. E. Strom: ApJ **427**, 961 (1994)
19. W. D. Heintz: J. Roy. Astron. Soc. Canada **63 (6)**, 275 (1969)
20. K.-W. Hodapp, E. F. Ladd: ApJ **453**, 715 (1995)
21. L. B. G. Knee, G. Sandell: A&A **361**, 671 (2000)
22. C. D. Koresko, T. M. Herbst, Ch. Leinert: ApJ **480**, 741 (1997)
23. P. Kroupa: MNRAS **277**, 1491 (1995)
24. R. B. Larson: MNRAS **272**, 213 (1995)
25. Ch. Leinert, H. Zinnecker, N. Weitzel et al.: A&A **278**, 129 (1993)
26. Ch. Leinert, A. Richichi, M. Haas: A&A **318**, 472 (1997)
27. O. Malkov, H. Zinnecker: MNRAS **321**, 149 (2001)
28. B. D. Mason et al.: 'First Results of a Speckle Survey of O Stars'. In: *The origins, evolution, and destinies of binary stars in clusters, Astronomical Society of the Pacific Conference Series, Volume 90, An international symposium held at the University of Calgary, 18-23 June 1995*, ed. by E. F. Milone, J.-C. Mermilliod (Astronomical Society of the Pacific, ASP, San Francisco 1996), p. 40
29. R. D. Mathieu: ARA&A **32**, 465 (1994)
30. M. J. McCaughrean, H. Zinnecker: in prep.
31. A. Moneti, H. Zinnecker: A&A **242**, 428 (1991)
32. T. Mouschovias: ApJ **211**, 147 (1977)
33. F. Palla, S. W. Stahler: ApJ **418**, 414 (1993)
34. Ch. F. Prosser, J. R. Stauffer, L. Hartmann et al.: ApJ **421**, 517 (1994)
35. B. Reipurth, H. Zinnecker: A&A **278**, 81 (1993)
36. F. H. Shu, F. C. Adams, S. Lizano: ARA&A **25**, 23 (1987)
37. M. Simon, A. M. Ghez, Ch. Leinert: ApJ **408**, L33 (1993)
38. M. Simon, A. M. Ghez, Ch. Leinert: ApJ **443**, 625 (1995)
39. H. Zinnecker: MNRAS **210**, 43 (1984)
40. H. Zinnecker: 'Pre-Main Sequence Binaries'. In: *Low Mass Star Formation and Pre-Main Sequence Objects, ESO Workshop in Garching, 11-13 July 1989*, ed. by B. Reipurth, (ESO Conference and Workshop Proceedings No. 33), p. 447
41. H. Zinnecker, W. Brandner, B. Reipurth: In: *Complemantary Approaches to Double and Multiple Star Research(1992) Astronomical Society of the Pacific Conference Series, Volume 32, Meeting held at Pine Mountain, Georgia, 5-10 April 1992*, ed. by H. A. McAlister, W. I. Hartkopf (Astronomical Society of the Pacific, ASP, San Francisco 1992), p. 50

42. H. Zinnecker, B. Wilking: 'Infrared Companions - Clues to Binary Star Formation'. In: *Binaries as Tracers of Stellar Formation, Proceedings of a Workshop held in Bettmeralp, Switzerland, Sept. 1991*, ed. by A. Duquennoy, M. Mayor (Cambridge University Press, Cambridge, New York 1992) p. 269

8 Low-Mass Binary Companions to Intermediate-Mass Stars in the Sco-Cen OB Association

8.1 Abstract

We propose to use ADONIS to search for close ($\geq 0.1''$) low-mass companions to the young A and B stars in the Upper Scorpius subgroup of the Scorpius-Centaurus OB association (distance 145 pc, age 5 Myr), for 3 reasons: 1) We expect to find faint (K = 10) companions of T Tauri type, which may account for the unexpected X-ray emission from some of these systems. 2) We will probe a regime in brightness ratio unaccessible but complementary to Hipparcos astrometric binaries and existing spectroscopic surveys. 3) We will obtain the first systematic multiplicity study of early type stars with a substantial range in companion mass and semi-major axis, important for star formation theories. This is an ideal project for the adaptive optics system ADONIS, as the A and B stars are bright and can be used for high-order wavefront sensing.

8.2 Description of the Proposed Programme

Scientific Rationale. The companion frequency of young stars of all spectral types is currently a topic of considerable astrophysical interest. The accurate characterisation of this frequency as a function of mass, separation and environment provides crucial constraints for star formation theories. We have been engaged in NIR speckle surveys of the multiplicity of young low-mass stars (T Tauri stars of spectral type G and K) in the Taurus T association [15]. Subsequently, the multiplicity of X-ray (ROSAT) selected young weak-line T Tauri stars in the Sco-Cen OB association was studied in the same way. Meanwhile 100 stars have been surveyed with SHARP at the ESO-NTT and detailed analysis indicates [13], [5] and [6] that the binary frequency is of the order of 85 %, somewhat higher than that of the G-dwarfs in the solar neighborhood [11].

Unfortunately, not much is known about the companion frequency of the more massive stars in OB associations like Sco-Cen or Orion. High spatial resolution bispectrum observations and multiplicity results of 13 Orion OB stars were recently reported [22] and [18], however, Sco-Cen is better suited for an investigation, as it is 3 times closer than Orion, at a distance of only 145 pc. Searches for spectroscopic binaries among massive and intermediate mass stars exist, but the majority of companions lies at wider separations (median of the order of 30 AU or 0.2 arcsec in Sco-Cen). The scientific interest in binarity studies of higher mass stars derives from the question whether the different processes of high-mass and low-mass star formation [2] reflect themselves in different binary frequencies and distribution of mass ratios [3].

Immediate Objective

1) We want to observe all the young A and B stars known to be members of the Sco-Cen OB association (*cf.* Fig. 24). In this proposal we will focus on the A and B stars in the Upper Scorpius subgroup of the association. There is a total of about 85 objects in this subgroup [10] for which we will establish whether they have close visual companions of the T Tauri type in the range of projected separation between about 0.1 arcsec (the diffraction limit of the 3.6 m at 2.2 μm) to a few arcsec. With a sample of 85 early-type stars we expect to obtain statistical information on *the companion frequency in the nearest complete sample of young A and B stars.* These stars, being more massive than solar-mass stars, also allow us to probe a larger range in component mass ratio. For example, our observations could reveal first hints whether the A stars typically have fainter (*i.e.* lower mass) companions than the B stars.

2) Another goal of the present observations will be whether we can detect companions to those A and B stars which ROSAT has found to be X-ray emitters [1] (these authors show that at least 20 % of A and B stars in Sco-Cen have X-ray emission characteristic of T Tauri stars). Here we must note that neither A stars nor B2 - B9 stars are supposed to be X-ray sources, since stars in that spectral range have neither strong winds nor coronae. In order to detect these suspected X-ray emitting T Tauri like companions, optical speckle methods are not suitable, as the magnitude differences between the bright early-type primaries and the faint late-type secondaries in the visual is rather extreme, *i.e.* of the order of 8 - 10 magnitudes. However, in the near-infrared (J,H,K) the contrast between the primary and secondary components is much more favourable, *i.e.* only 4 - 5 magnitudes difference. This much improved contrast in the infrared and the fact that the primaries are very bright ($V = 4^m - 7^m$) visual objects, ideal for high order wavefront sensing, makes this a perfect project for ESO's state of the art ADONIS system.

3) We would like to stress here that present spectroscopic studies of binarity [16] and [21] are not sensitive to low-mass companions around A and B stars, for separations of 0.1 arcsec or more. Similarly, Hipparcos astrometric studies are also insensitive to these types of binaries with very large brightness differences at the distance of Sco-Cen. Hipparcos observations are sensitive down to separations of 0.1 arcsec for magnitude differences (in the visual) of

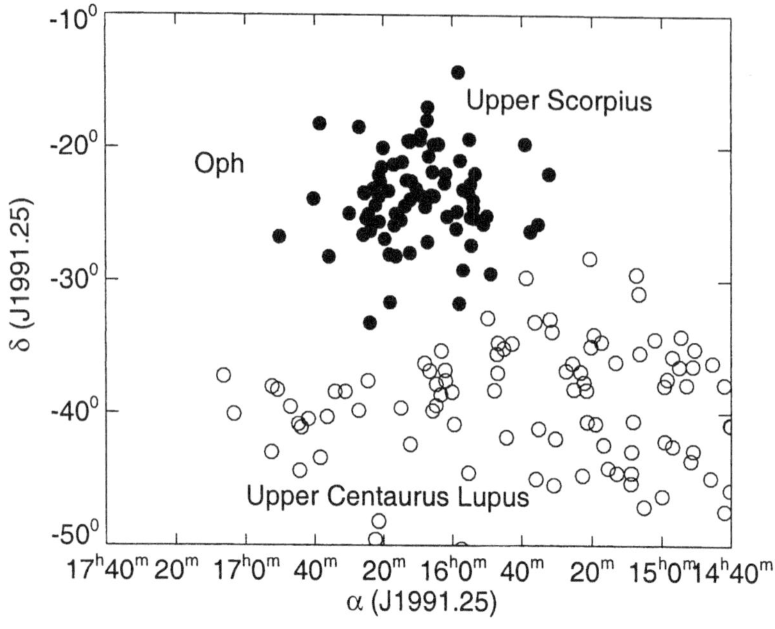

Fig. 24. Spatial Distributions of Hipparcos selected A and B stars in the Sco-Cen OB association

less than about 3 magnitudes. Our proposed study thus fills the gap between present spectroscopic and astrometric studies of binarity.

All our targets have been observed by Hipparcos and the majority of them are confirmed as members of the Upper Scorpius subgroup of Sco-Cen, based on both proper motion and parallax data [10]. To complement our adaptive optics study further information on multiplicity is available from the Hipparcos catalogue and from the literature, which one of us has recently extensively culled. Furthermore, two follow-up studies have been done for Upper Scorpius based on astrometric data. In the first of these, improved estimates of the Hipparcos parallaxes were used to construct an accurate Hertzsprung-Russell diagram [9]. The second study [12] consists of an extensive comparison between Hipparcos proper motions and those in the ACT [20] and TRC [14] catalogues. Both studies allow the identification of additional candidate binaries as well as possible members missed by the Hipparcos study. This extends further the complementary information on binarity and provides a number of additional targets.

Finally, we also aim at improving the absolute photometric calibration of our stars, at least to a level of 30 %, made possible by the new parallaxes (1 mas rms for a parallax of 6.9 mas or 145 pc corresponds to an uncertainty in distance of about 15 %). Many astrophysical problems hinge on a good absolute brightness calibration of young intermediate mass stars [19] which in return will lead to more accurate IMF determinations and to improved stellar evolutionary models. Our studies will also improve knowledge of the basic stellar quantities (T_{eff}, radius, log g, mass). Hence, the proposed observations address fundamental questions related to stellar evolution and star formation.

8.3 Results

Adaptive optics K-band observations searching the immediate neigbourhood of 200 A and B stars in the Scorpius-Centaurus OB association, with Hipparcos confirmed membership, have found 67 "companions" in the range 30 to 300 AU. Many of these may not constitute physical systems but optical pairs instead (see Shatsky and Tokovinin astro-ph/0109456 for a similar study). Nevertheless the implied total binary frequency of A and B stars appears to be high (more than 50 %) and the implied mass ratios inferred from the K-band brightness ratios suggest that A and B stars prefer F and G stars as companions rather than K and M stars. Thus the concept of random pairing from a pre-existing IMF (Malkov and Zinnecker 2001, MNRAS **321**, 149) seems to be wrong, *i.e.* the secondary component at the time of formation seems to "know" something about the mass of its primary component.

References

1. T.W. Berghöfer, J.H.M.M. Schmitt, J.P. Cassinelli: A&AS **118**, 481 (1996)
2. I.A. Bonnell: NATO Science Ser. Vol. **540**, p479 (1999)
3. I.A. Bonnell, M.R. Bate, H. Zinnecker: MNRAS **298**, 93 (1998)
4. W. Brandner, H. Zinnecker: A&A **321**, 220 (1997)
5. W. Brandner, J.M. Alcala, M. Kunkel et al.: A&A **307**, 121 (1996)
6. W. Brandner, R. Köhler: ApJ **499**, L79 (1996)
7. A.G.A. Brown, W. Verschueren: A&A **319**, 811 (1997)
8. A.G.A. Brown, A. Blaauw, R. Hoogerwerf, J.H.J. de Bruijne, P.T. de Zeeuw: NATO Science Ser. Vol. **50**, p411 (1999)
9. J.H.J. de Bruijne,: MNRAS, **310**, 585 (1999)
10. P.T. de Zeeuw, R. Hoogerwerf, J.H.J. de Bruijne, A.G.A. Brown, A. Blaauw: AJ **117**, 354 (1999)
11. A. Duquennoy, M. Mayor: A&A **248**, 485 (1991)
12. R. Hoogerwerf: MNRAS, **313**, 43 (2000)
13. R. Köhler, M. Kunkel, C. Leinert, H. Zinnecker: A&A, **356**, 541 (2000)
14. A. Kuzmin, G. Høe, U. Bastian et al.: A&AS **136**, 491 (1999)
15. Ch. Leinert, H. Zinnecker, N. Weitzel et al.: A&A **278**, 129 (1993)
16. H. Levato, S. Malaroda, N. Morrell, G. Solivella: ApJS **64**, 487 (1987)
17. T. Preibisch, H. Zinnecker: AJ **117**, 2381 (1999)
18. T. Preibisch, Y. Balega, K.-H. Hofmann, G. Weigelt, H. Zinnecker: NewA **4**, 531 (1999)
19. D. Schoenberner, P. Harmanec: A&A **294**, 509 (1995)
20. S.E. Urban, T.E. Corbin, G.L. Wycoff: AJ **115**, 2161 (1998)

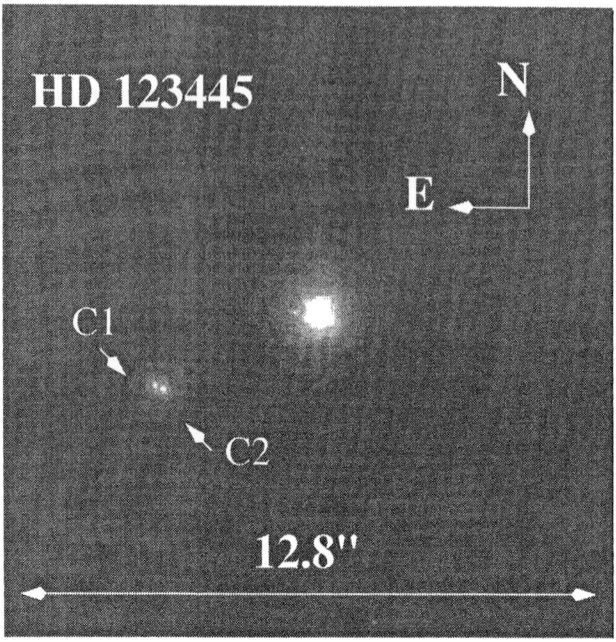

Fig. 25. K-band image of HD 123445 (A B9V star in the Upper Centaurus Lupus association), observed with the adaptive optic system ADONIS at the ESO 3.6 m telescope. The whole field of the SHARP II camera (12.8 arcsec × 12.8 arcsec) is shown. As can be seen, two new objects are detected at ∼ 5 arcsec NE from the central star (perhaps a true 0.25 arcsec binary system). Courtesy of N. Huélamo; see: N. Huélamo, W. Brandner, A. G. A. Brown, R. Neuhäuser, and H. Zinnecker: 'ADONIS observations of hard X-ray emitting late B-type stars', In: A&A **373**, 657 (2001)

21. W. Verschueren, M. David, A. G. A. Brown: 'A Search for Early-Type Binaries in the SCO OB2 Association'. In: *The Origins, Evolution and Destinies of Binary Stars in Clusters, Astronomical Society of the Pacific Conference Series, Volume 90, An international symposium held at the University of Calgary, 18-23 June 1995*, ed. by E. F. Milone, J.-C. Mermilliod (Astronomical Society of the Pacific, ASP, San Francisco 1996) p. 131
22. G. Weigelt, Y. Balega, Th. Preibisch et al.: A&A **347**, L15 (1999)

9 Multiplicity of Massive Stars in the Orion Nebula Cluster and Implications on Their Formation Mechanism**

9.1 Abstract

We present the results of a bispectrum speckle interferometric survey for binaries among the massive stars in the Orion Nebula cluster. Observations of 13 bright cluster members of spectral type O or B reveal 8 visual companions in total. Using the flux ratios of the resolved systems to estimate the masses of the companions, we find that the systems generally have mass ratios below 0.5. Extrapolation with correction for the unresolved systems suggests that there are at least 1.5 companions per primary star on average. This number is clearly higher than the corresponding number for low-mass primaries, suggesting that a different mechanism is at work in the formation of high-mass multiple systems than for low-mass multiple systems.

9.2 Introduction

The knowledge of the multiplicity of very young massive stars can provide important information on their formation mechanism, which is still not well understood (*cf.* [17]). High-mass stars probably cannot form through gravitational collapse in molecular cloud cores and subsequent accretion like the low-mass stars, because as soon as the stellar core reaches a mass of $\sim 10\,M_{\odot}$, the radiation pressure on the infalling dust halts the accretion and thus limits the mass [20]. Bonnell et al. [3] suggested that high-mass stars form through accretion-induced collisions of protostars in the dense central regions of forming stellar clusters. Their theory predicts that multiple systems should be very common amongst the massive stars, due to frequent tidal captures.

We have performed a survey for visual binaries in massive stars in the very young Orion Nebula cluster (ONC, *cf.* [6]; [7]; [8]), which can help to test this prediction. For spectroscopic binaries among massive stars see [22].

The alternative models of massive star formation (the so-called "strong accretion" model pursued by Maeder and coworkers [2], [12]; and the disk accretion model by Yorke [21]) have not yet predicted the binary properties of massive stars.

9.3 Observations

Our sample consists of 13 bright ONC members with spectral types O or B, all located within 20 arcmin of the Trapezium (*cf.* Table 1). The speckle interferograms were obtained with the 6 m telescope at the Special Astrophysical Observatory (SAO) in Russia in October 1997. We obtained data in the near-infrared H- and K-bands. The diffraction-limited images were reconstructed using the bispectrum speckle interferometry method [18]; [11]; [9].

** In collaboration with Th. Preibisch and G. Weigelt (see IAU-Symp. 200)

Table 1. Target stars of our survey.

Par #	Brun #	other name	spectral type	V [mag]
1605	388	V372 Ori	B9.5+A0.5	7.98
1744	502	HD 36981	B5	7.89
1772	530	LP Ori	B1.5	8.47
1863	595	θ^1 Ori B	B2	7.96
1865	587	θ^1 Ori A	B0	6.73
1889	612	θ^1 Ori D	B0.5	6.71
1891	598	θ^1 Ori C	O6	5.12
1993	682	θ^2 Ori A	O9.5	5.07
2031	714	θ^2 Ori B	B1	6.41
2074	747	NU Ori	B1	6.84
2271	907	HD 37115	B6	7.00
2366	980	HD 37150	B3	6.56
2425	1018	WH 349	B6	10.64

9.4 Results

In our speckle reconstructions we were able to resolve visual companions of 7 of the 13 target stars. Figure 26 shows our reconstructed images of the resolved systems. Interestingly, in all resolved systems the K-band flux ratios are quite small, *i.e.* less than 1/3. This suggests that all these visual companions are much less luminous and thus less massive than the primary stars.

Estimation of Companion Masses. We have estimated the masses of the companion stars in the resolved systems from the observed K- and H-band flux ratios in our images. These flux ratios, together with the assumption that the companions are very young pre-main sequence stars, can be used to estimate their location in the HRD, allowing us to determine stellar masses by comparison with pre-main sequence tracks. All details of this method can be found in [19] and [14]. Since the masses estimated in this way might be subject to significant uncertainties, we have also determined firm upper limits for the companion masses by assuming the companions to be main-sequence stars. From the companion masses we have computed the mass-ratios $q := M_c/M_p$, which are listed in Table 2; the upper limits to the mass-ratios are added in brackets.

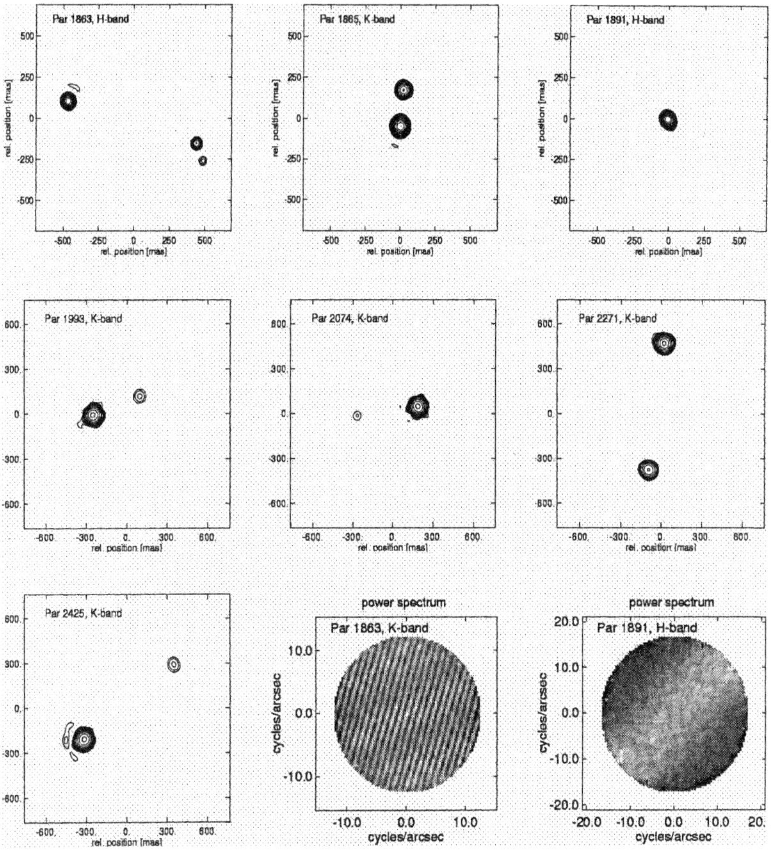

Fig. 26. Diffraction-limited images of the resolved systems, reconstructed by the bispectrum speckle interferometry method. The contour level intervals are 0.3 mag, down to 3.6 mag difference relative to the peak intensity. North is up and east is to the left. The two plots in the lowest row show reconstructed power spectra of Par 1863 and Par 1891.

The Distribution of Mass Ratios. In Fig. 27 we compare our empirical mass ratio distribution for the ONC stars to several different model distributions: **(a)** to a companion mass function corresponding to the [15] field IMF, **(b)** to a flat companion mass function (i.e. all companion masses are equally likely), **(c)** to a distribution that slightly favours massive companions, and **(d)** to a distribution that strongly favours systems with (nearly) equal masses. All details about this distributions can be found in [14].

The observed distribution function for the stars in our sample systematically lies above the simulated distribution functions for models (b), (c), and (d). If we take into account our observational bias against systems with low mass ratios, which are hard to detect due to contrast problems, the true mass

Table 2. Summary of all known companions of the observed stars. References: 1: [14]; 2: [19]; 3: [13]; 4: [4]; 5: [16]; 6: [1]; 7: [10]

Prim. Par	M_p [$M_\odot$]	Comp	ρ [AU]	q	Ref
1605-1	3.5	-2 (spec)		$\sim 0.9-1.0$	7
1863-1	7	-2 (vis)	430	$\sim 0.22\,(<0.71)$	1,2
		-3 (vis)	460	$\sim 0.10\,(<0.50)$	1,2
		-4 (vis)	260	$\sim 0.03\,(<0.29)$	1,5
		-5 (spec)	0.13		6
1865-1	16	-2 (vis)	100	~ 0.25	2,3
		-3 (spec)	1	~ 0.13	4
1891-1	45	-2 (vis)	16	~ 0.12	2
1993-1	25	-2 (vis)	173	$\sim 0.28\,(<0.32)$	1
		-3 (spec)	0.47	~ 0.35	6
2074-1	14	-2 (vis)	214	$\sim 0.07\,(<0.28)$	1
		-3 (spec)	0.35	~ 0.2	6
2271-1	5	-2 (vis)	400	$\sim 0.29\,(<0.96)$	1
2425-1	4	-2 (vis)	388	$\sim 0.04\,(<0.35)$	1

ratio distribution must be even steeper in the low-q part. Thus it seems very likely that the true distribution of mass ratios in our sample is not consistent with any of the models (b), (c), or (d), but might in principle be consistent with the field IMF model (a).

The Binary Frequency. It is obvious that our data do not allow us to detect all multiple systems; we can only see those systems which are wide enough and for which the flux ratio is not too low. In order to correct for this effect, we have performed numerous simulations of multiple systems to estimate the fraction of undetected binary systems. We assume that the distribution of orbital periods is the same as for solar-mass primaries found by [5]. For the companions we consider either a mass function according the the [15] field IMF (model a), or the flat mass distribution (model b). All details of our simulations are described in [14].

The simulation results for the case of the [15] field IMF are shown in Fig. 28. In this figure we have marked the region in the flux-ratio versus separation space, which contains those simulated binary system we would

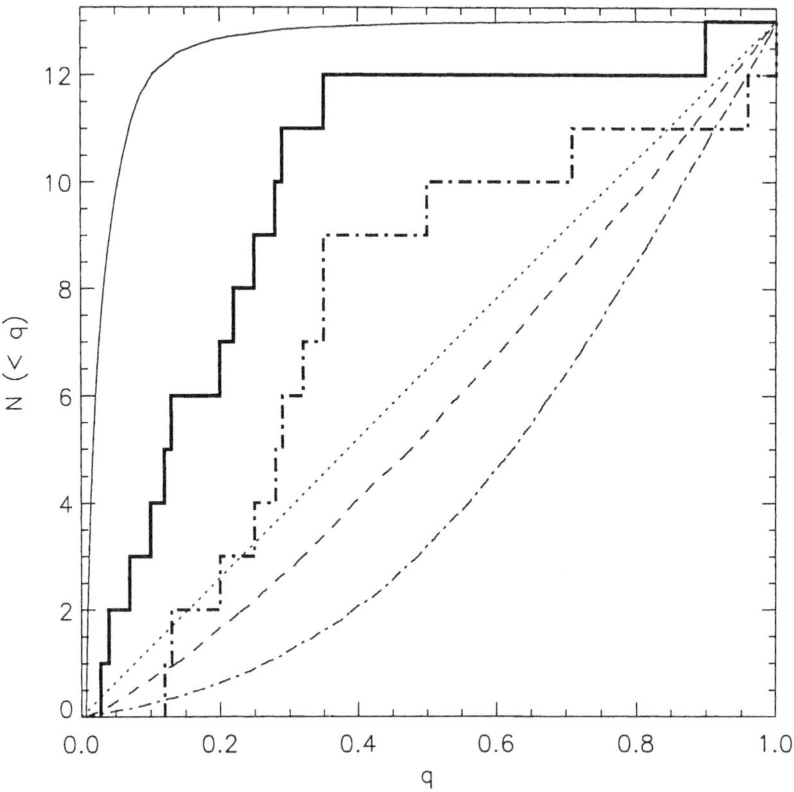

Fig. 27. Distribution function of observed mass ratios of the ONC multiple systems (thick solid line) compared to the four model functions discussed in the text: (a): Scalo (1998) field IMF (thin solid line); (b): flat mass ratio distribution (thin dotted line); (c): massive companions slightly favoured (dashed line); (d): systems with equal masses strongly favoured (thin dashed-dotted line). The thick dashed-dotted line shows the observed distribution function based on the upper mass limits for the companions.

be able to resolve in our data. The boundaries of this area are defined by our minimal detectable separation of ~ 33 mas, the maximum separation of 1.75 arcsec (given by the 3.5 arcsec $\times$ 3.5 arcsec field of our images which we inspected for companions), and the minimum detectable flux-ratio as a function of the separation. The fraction of binaries which would be detectable in our data can now easily be determined by counting the number of simulated systems within this area. We find that the detectable fractions are 15 % and 40 % for the [15] field IMF and the flat mass ratio distribution, respectively. Thus, we conclude that a detectable fraction of ~ 40 % certainly is a con-

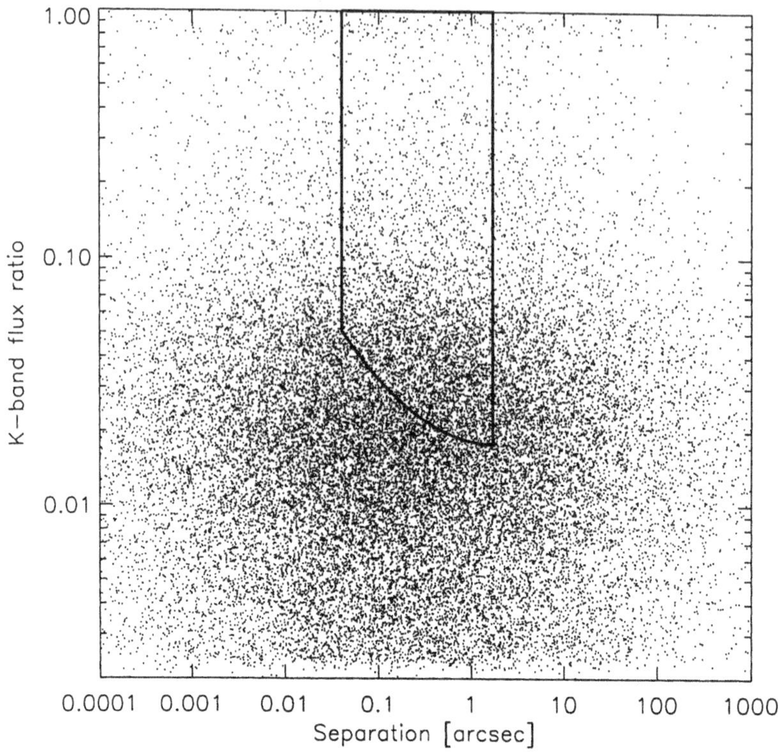

Fig. 28. Results of the simulations for the case of companions from the Scalo (1998) field IMF (model a). Each simulated system is shown as a dot in the separation – flux ratio space. The area delimited by the thick line marks the region of systems detectable in our data.

servative upper limit, *i.e.* that the correction factor to extrapolate from the detected companions to the full population of companions is at least ~ 2.5.

Since we find visual companions to 7 of the 13 target stars, *i.e.* see an apparent binary frequency of $(54 \pm 14)\,\%$, the true binary frequency must be very close to 100 %. Given the number of 8 detected visual companions, the true number of companions is ≥ 20, suggesting a mean number of ≥ 1.5 companions per primary star. This is clearly higher than the corresponding number for the low-mass stars in the Orion nebula cluster and the field star population (~ 0.5 companions per primary star on average; *cf.* [5]; [13]; [16]).

Multiplicity Versus Spectral Type. A trend for a higher degree of multiplicity among the stars of very early spectral type as compared to the later type stars is apparent in our sample, as can be seen in Fig. 29. The average number of known (visual & spectroscopic) companions per primary is 2.3 times higher among the primaries with spectral type earlier than B3 (11

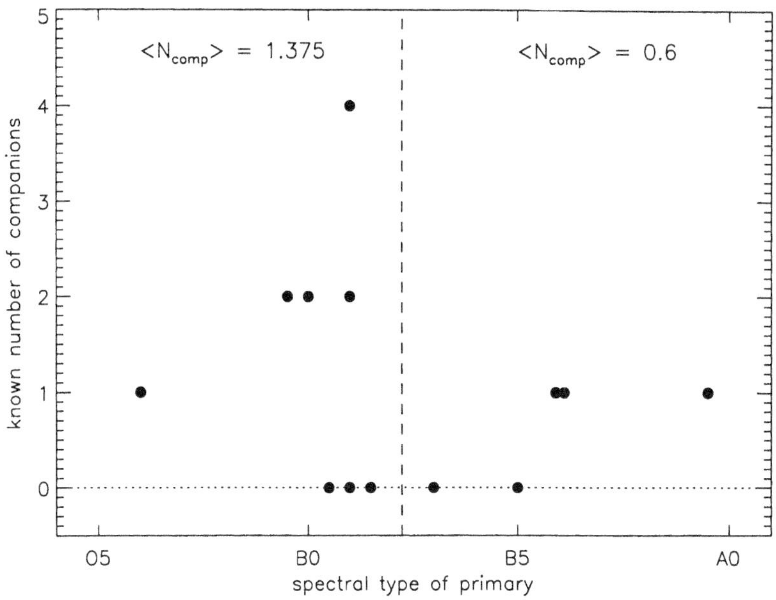

Fig. 29. The known number of companions is plotted against the spectral type of the primary.

known companions to 8 primaries) than among the later type primaries (3 known companions to 5 primaries).

9.5 Conclusions

Our data show that the multiplicity of the massive stars in the ONC is quite high (≥ 1.5 companions per primary star on average, after correction for unresolved systems) and significantly ($\sim 3\times$) higher than among low-mass stars. The stars of spectral type earlier than B3 display a higher degree of multiplicity than the later type stars. These findings suggests different formation mechanisms for the high-mass and low-mass multiple systems. The nature of our results seems to support the idea that high-mass ($\mathrm{M} > 10\,M_\odot$) stars form through collisions of protostars.

References

1. H. A. Abt, R. Wang, O. Cardona: ApJ **367**, 155 (1991)
2. R. Behrend, A. Maeder: A&A **373**, 190 (2001)
3. I. A. Bonnell, M. R. Bate, H. Zinnecker: MNRAS **298**, 93 (1998)
4. M. Bossi, A. Gaspani, M. Scardia, M. Tadini: A&A **222**, 117 (1989)
5. A. Duquennoy, M. Mayor: A&A **248**, 485 (1991)

6. G. H. Herbig, D. M. Terndrup: ApJ**307**, 609 (1986)
7. L. A. Hillenbrand: AJ **113**, 1733 (1997)
8. L. A. Hillenbrand, L. W. Hartmann: ApJ **492**, 540 (1998)
9. K.-H. Hofmann, W. Seggewiss, G. Weigelt: A&A **300**, 403 (1995)
10. H. Levato, H. A. Abt: PASP **88**, 712 (1976)
11. A. W. Lohmann, G. Weigelt, B. Wirnitzer: Appl. Opt. **22**, 4028 (1983)
12. P. Norberg, A. Maeder: A&A **359**, 1025 (2000)
13. M. G. Petr, V. Du Foresto, S. V. M. Beckwith: ApJ **500**, 825 (1998)
14. Th. Preibisch, Y. Balega, K.-H. Hofmann, G. Weigelt, H. Zinnecker: New Astronomy **4**, 531 (1999)
15. J. Scalo: 'The IMF Revisited: A Case for Variations'. In: *The Stellar Initial Mass Function, Astronomical Society of the Pacific Conference Series, Volume 142, 38th Herstmonceux Conference*, ed. by G. Gilmore, D. Howell,(1998) p. 201
16. M. Simon, L. M. Close, T. L. Beck: AJ **117**, 1375 (1999)
17. S. W. Stahler, F. Palla, P. T. P. Ho: 'The Formation of Massive Stars'. In: *Protostars and Planets IV*, ed. by V. Mannings, A. P. Boss, S. S. Russell, (University of Arizona Press, Tucson, 2000) p. 327
18. G. Weigelt: Opt. Commun. **21**, 55 (1977)
19. G. Weigelt, Y. Balega, Th. Preibisch, D. Schertl, M. Schöller, H. Zinnecker: A&A **347**, L15 (1999)
20. H. W. Yorke, E. Krügel: A&A **54**, 183 (1977)
21. H. W. Yorke: 'Formation of Massive Stars by Accretion' in *The earliest phases of massive starbirth*, Proc. Hot Star Workshop III, Boulder, Colorado (2001), ed. P. Crowther
22. B. Garcia, J.C. Mermilliod, A&A, **368**, 122 (2001)

IV. Low Mass Stellar Content of Young Star Clusters and Associations

10 The Low-Mass Initial Mass Function in the Upper Scorpius OB Association

10.1 Abstract

We propose to use the 2dF multi-fiber spectrograph to obtain a complete sample of low-mass pre-main sequence (PMS) stars in a 12 square-degree area of the Upper Scorpius OB association. From previous studies and from our recent multi-object spectroscopic observations with FLAIR we know already about 100 PMS stars in Upper Sco. However, most of the late K- and M-type stars were too faint for FLAIR and thus are missing in our sample. From SuperCOSMOS measurements we have selected some 1400 candidates for low-mass PMS stars from a colour-magnitude diagram. Spectroscopy of these stars with 2dF will allow us to identify PMS stars by their Li 6708 Å absorption lines. We will be able to determine the (0.1 - 20 $M_\odot$) IMF in an OB association for the first time.

10.2 Scientific Justification

We ask for another chance to use the unique capabilities of 2dF to reveal the full low-mass stellar population in the Upper Scorpius OB association (during our observing run this June we lost 75 % of our time due to bad weather). The population of low-mass ($M_\star \leq 2\,M_\odot$) pre-main sequence (PMS) stars in OB associations is important in order to derive the Initial Mass Function (IMF) over the full mass range, to understand their star formation history, and to compare these properties to those of low-mass star forming regions like T associations. In contrast to the rather quiescent conditions in T associations, the conditions in OB association are characterized by strong winds of the luminous O and B stars and frequent supernova explosions. These differences might affect the mass functions, for example by making it harder for low-mass stars to form. However, the low-mass IMF of OB associations is still very poorly known, because low-mass PMS stars in OB associations are very difficult to find: The only completely reliable sign for their PMS nature is the strength of the 6708 Å Lithium line, requiring at least intermediate resolution spectroscopy for accurate measurement. Thus, the huge observational effort to identify the widespread population of PMS stars among the many thousands of field stars usually prevents further investigations. The unique capabilities of 2dF now allow such an investigation for the first time. The best OB association to study is Upper Scorpius (*cf.* [1]; [11]; [14]), because it is the nearest OB association (D = 145 pc) and is essentially free of extinction ($A_V \leq 2$ mag). Upper Sco was also well studied by Hipparcos, and thus

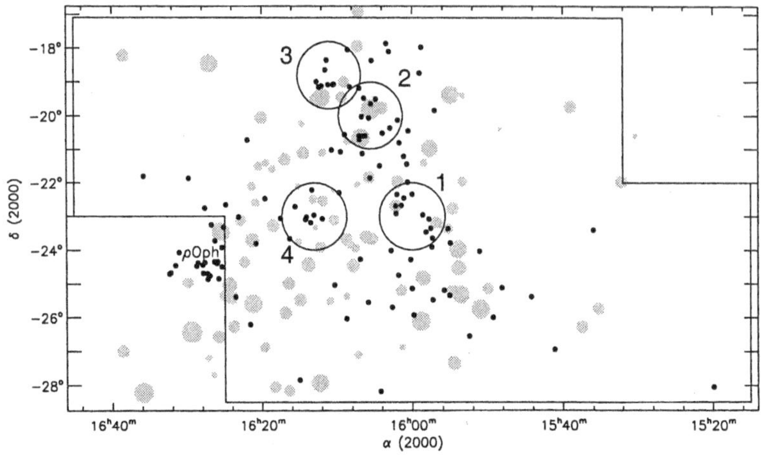

Fig. 30. Map of the Upper Scorpius region. The thick solid line marks the 160 square-degree area we have surveyed with FLAIR. The low-mass PMS stars are shown as small solid dots. The Hipparcos high-mass members are shown as big grey dots (symbol size proportional to $\log L_\star$). The circles show the proposed 2dF fields.

the population of high- and intermediate-mass ($M_\star \geq 3\,M_\odot$) stars is completely known [14]. We have recently performed a spectroscopic survey for PMS stars in a 160 square-degree field in Upper Sco. Using the multi-object spectrograph FLAIR at the UKST, we could identify 39 new PMS stars [9]. In combination with two earlier studies [13]; [4], our current sample contains 98 low-mass PMS stars. We have constructed a new HR diagram and found that the age of these stars is about 5 Myr, with relatively small dispersion, and that it agrees very well with the age of the B stars [11]. While FLAIR was an excellent tool to cover the huge area of the association, it did not allow us to investigate – with the required spectral resolution – the fainter late K- and M-type stars. Thus, our current sample is only complete for the rather bright stars with $M_\star \geq 0.8\,M_\odot$. From SuperCOSMOS measurements of the R and I UKST survey plates of our region we have constructed a colour-magnitude diagram and selected candidates for PMS stars (Fig. 31). Since this photometrically selected sample is contaminated by field stars, spectroscopy is necessary to confirm the PMS nature of these candidates and to allow accurate stellar masses to be determined. We want to characterize the mass function by comparing the numbers of low-mass stars to those of the high-mass stars. According to the Miller-Scalo field IMF, there should be 11.5 times more stars with 0.3 - 1.0 $M_\odot$ and 17.2 times more stars with 0.1 - 0.3 $M_\odot$ than stars with $M \geq 3\,M_\odot$. If we want to be able to draw strong conclusions about the mass function in Upper Sco, we have to cover an area which is large enough that the expected number of low-mass stars is

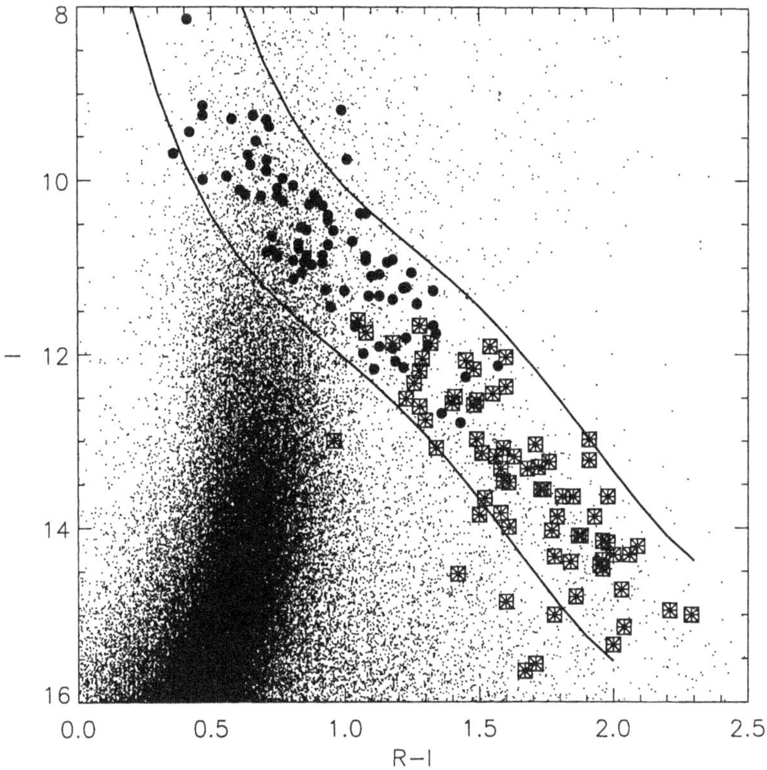

Fig. 31. Colour-Magnitude diagram for the four 2dF fields. The thick solid dots show the previously known PMS stars in Upper Sco. The small dots show the SuperCOSMOS data for all stars in the four 2dF fields. The solid lines mark the region of photometric candidates for PMS stars (based on the theoretical 5 Myr isochrone). In order to take into account photometric errors and a possible age spread, candidates have been selected from a somewhat wider band. The new PMS stars detected in our 2dF observations of field Upper Sco 2 are shown as asterisks surrounded by squares.

at least 100 (only in that case will we be sensitive for 20 % deviations at a confidence level of 0.95). Since the average number of stars with $\mathrm{M} \geq 3\,M_{\odot}$ in Upper Sco is about 0.6 per square-degree, we have to cover an area of at least 12 square-degrees (then we can expect $0.6 \times 12 \times 17.2 \approx 120$ stars with $0.1 - 0.3\,M_{\odot}$). A 12 square-degree area will contain about 1400 photometric PMS candidates and is also large enough to avoid selection effects due to poor spatial coverage. From these numbers, it is clear that 2dF is the only instrument which makes it possible to reach these goals. We have selected four fields (Fig. 30), containing a total of 1381 PMS candidates (Fig. 31). We will use the 2dF spectra of these stars to measure the width of the 6708 Å

Lithium line and to determine the spectral type. This will provide definitive evidence on the PMS nature of these stars. We then can determine the masses and ages of the new PMS stars from their position in the HR-diagram by comparison with theoretical PMS tracks (*e.g.* [7]). If the mass function is the same as the field IMF, we expect to find about 120 stars with 0.1 - 0.3 $M_\odot$ and about 80 stars with 0.3 - 1.0 $M_\odot$. It will also be interesting to compare the mass function to the results from T associations like Taurus-Auriga or young clusters like IC 348 (*cf.* [8]; [3]). Finally, we will also be able to look for possible effects of the immediate stellar environment, since two of our fields contain several luminous B stars, while the two other fields contain no B stars. In the spectra of the field we could observe up to now (Upper Sco 2) we could detect already 77 new PMS stars (cf. Fig. 31). This demonstrates that our project is feasible and very successful. It also suggests that the number of low-mass PMS stars might be even somewhat higher than expected from the field IMF. To put this very interesting result on a reliable statistical basis, it is very important to observe the other 3 fields too. Only then we will be able to draw strong conclusions.

Acknowledgments. This as well as the following proposal was written by Th. Preibisch (PI) and HZ.

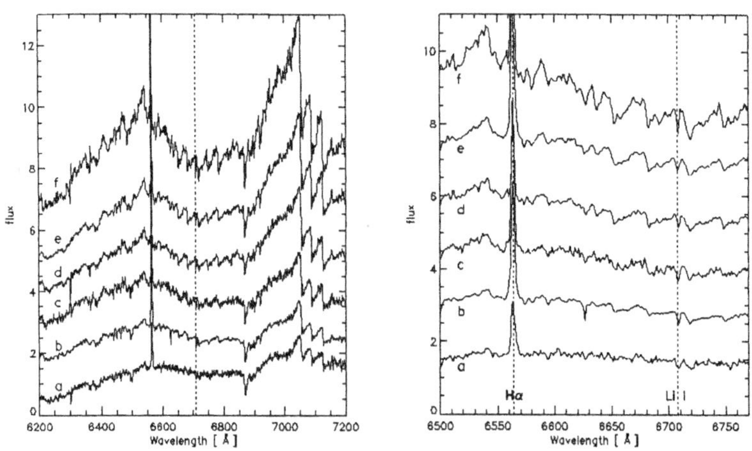

Fig. 32. Typical examples of our 2 dF spectra. Spectra of the following PMS stars are shown: (a), 160954.4 – 190654 (M1); (b), 160502.1 – 203507 (M2); (c), 160823.5 – 191131 (M3); (d), 160516.1 – 193830 (M4); (e), 160350.4 – 194121 (M4.5); and (f), 160632.1 – 202053 (M5). The left plot shows (nearly) the full wavelength range of our spectra ; the right plot shows details in the region of the Hα and Li lines.

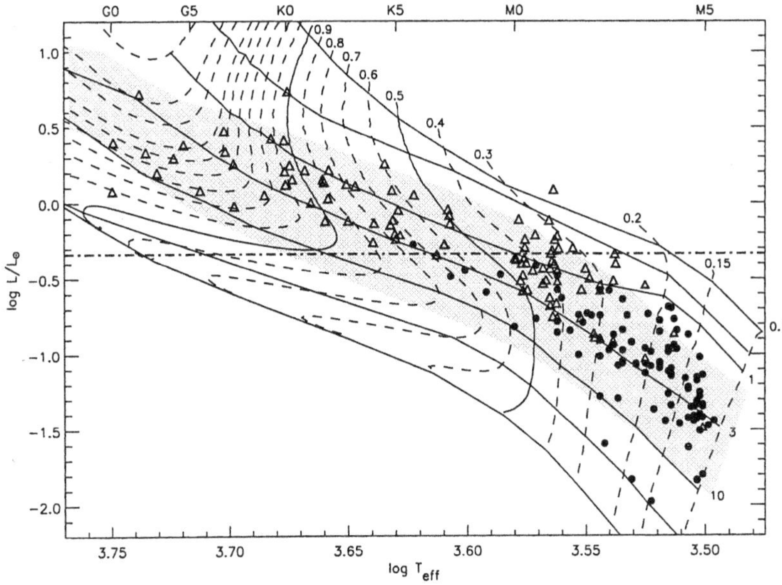

Fig. 33. H-R diagram for the PMS stars in Upper Sco. The previously known PMS stars from PZ 99 are shown as triangles, and the new PMS detected in this study are shown as circles. The dashed lines show the evolutionary tracks from D'Antona & Mazzitelli (1994) and are labeled by their masses in solar units. The 1 M⊙ and 0.5 M⊙ tracks are shown as solid lines. The other solid lines show the D'Antona & Mazzitelli (1994) isochrones for the ages of 0.1, 0.3, 1, 3, 10, and 30 Myr and finally the main sequence. The gray band shows the region in which we expect 90 PMS stars to lie, based on the assumption of a common age of 5 Myr for all stars and taking proper account of the uncertainties and the effects of unresolved binaries. The dash-dotted line indicates the sensitivity limit of the X-ray observations that were the basis of the X-ray selected PMS sample (see PZ 99).

10.3 Results

Utilizing the 2dF multiobject spectrograph at the Anglo-Australian Telescope, we obtained spectra of 576 stars with magnitudes $12.5 \leq R \leq 16.5$ in a 6 deg^2 area in Upper Scorpius, one of the subgroups of the nearest OB association. Among these objects we were able to identify 98 new pre-Main Sequence (PMS) stars, nearly all of them M-type stars, by their strong lithium absorption lines. We place the new PMS stars into the H-R diagram and find that their ages are around 5 Myr and their masses are fully consistent with the low-mass (0.1 to 0.5 $M_{\odot}$) field star initial mass function. Thus, the low-mass star are by and large coeval with the high-mass stars and there is no deficit of low-mass stars in Upper Scorpius. (See: Th. Preibisch, E. Guenther, H. Zinnecker: 'A large spectroscopic survey for young low-mass members in the Upper Scorpius OB association'. AJ **121**, 1040 (2001).

References

1. A. Blaauw: *The Physics of Star Formation and Early Stellar Evolution*, ed. by C.J. Lada and N.D. Kylafis, (1991) p. 125
2. E.J. de Geus, P.T. de Zeeuw, J. Lub: A&A **216**, 44 (1989)
3. G.H. Herbig: ApJ **497**, 736 (1998)
4. M. Kunkel: PhD Thesis, Universität Würzburg, Würzburg (1998)
5. G.E. Miller, J.M. Scalo: PASP **90**, 506 (1978)
6. G.E. Miller, J.M. Scalo: ApJS **41**, 513 (1979)
7. F. Palla, S.W. Stahler: ApJ **525**, 772 (1999)
8. Th. Preibisch, H. Zinnecker, G.H. Herbig: A&A **310**, 456 (1996)
9. Th. Preibisch, H. Zinnecker, E. Guenther et al.: A&A **333**, 619 (1998)
10. Th. Preibisch, H. Zinnecker, E. Guenther et al.: ASP Conference Series **154**, 1780 (1998)
11. Th. Preibisch, H. Zinnecker: AJ **117**, 2381 (1999) [PZ99]
12. Th. Preibisch, H. Zinnecker: 'Star formation in the Upper Scorpius OB association: evidence for a supernova-triggered mini-starburst'. In: *Stellar Clusters and Associations: Convection, Rotation, and Dynamos.*, ed. by R. Pallavicini, G. Micela, S. Sciortino, Proceedings from ASP Conference, **198**, 219 (2000)
13. F.M. Walter, F.J. Vrba, R.D. Mathieu, A. Brown, P.C. Myers: AJ **107**, 692 (1994)
14. P.T. de Zeeuw, R. Hoogerwerf, J.H.J. de Bruijne: AJ **117**, 354 (1999)

11 The Full Population of T Tauri Stars of the Young Stellar Cluster IC 348

11.1 Abstract

We propose to obtain a 50 ksec ACIS-I observation of the young stellar cluster IC 348. The high sensitivity and spatial resolution of AXAF/ACIS will allow us, for the first time, to detect X-ray emission from the numerous faint T Tauri stars in IC 348. This will enable us to identify the *full* population of T Tauri stars in IC 348, *i.e.* down to stellar masses of $0.1\,M_\odot$, by their X-ray emission.

Furthermore, this observation will provide us with X-ray spectra and lightcurves for a large (≥ 200) and homogeneous sample of T Tauri stars.

11.2 Scientific Justification

Background and Previous Observations. IC 348 is very young cluster located near the edge of the Per OB2 association at a distance of about 300 pc. Since the most massive star of the cluster (BD 31° 643) is only a B5V star, which is not powerful enough to disperse the cloud in which the stars were born (unlike an O star which disperses obscuring material more quickly), the cluster is still embedded in the remnant molecular core and most stars are obscured by 2 or more mag of visual extinction.

The cluster was long believed to consist of just a dozen T Tauri members [3]; [2]. However, recent sensitive observations have shown that it is much richer. Herbig [4] could detect over 110 Hα emission-line stars in the area, of which a majority are most likely T Tauri members of IC 348. A comparison of their locations in the HRD with pre-main sequence evolutionary tracks showed that most of these stars have ages in the range 0.7 - 3 Myr, with a mean of 1.3 Myr. Because there is a steep rise in the number of emission-line stars between W(Hα) = 10 Å, and the detection limit near W(Hα) = 3Å, there are probably many more T Tauri stars with weaker Hα emission in IC 348. For any study of global properties of this interesting cluster, *e.g.* the mass function or the star formation history, it is crucial to include these missing stars. However, due to their faintness, these stars are very hard to identify by optical techniques.

Lada & Lada [6] have analyzed near-infrared (JHK) observations of the cluster, and additionally we have recently performed a very deep K-band survey (completeness limit K$\sim$17). More than 1000 infrared sources could be detected in a 14×14 arcmin area (see Fig. 34). While these infrared data are sensitive enough to detect even the faintest cluster members, they provide no easy way to discern between members and background stars. We can only estimate by comparison with background fields that about 400 of these infrared sources probably are associated with IC 348. Again, this demonstrates that there are many more cluster members awaiting discovery.

X-ray observations have proven to be an excellent tool to find T Tauri stars, since the ratio $L_X/L_{\rm bol}$ is about 3 orders of magnitude larger for T Tauri stars than typical for field stars (*cf.* [5]; [7]). We have performed deep ROSAT observations of IC 348 and could detect more than 100 X-ray sources [10] (see Fig. 35). Many X-ray sources could be identified with Hα emission-line stars. Additionally, we identified 56 X-ray emitting stars which very likely are T Tauri stars lacking strong Hα emission. While we could detect most of the optically bright (V < 17) T Tauri stars in our ROSAT images, the data were not sensitive enough to detect X-rays from the fainter (V > 17) T Tauri stars in IC 348. Furthermore, due to the limited spatial resolution of ROSAT, our data suffer from some source confusion at the center of the cluster, where the source density is very high.

The minimum detectable flux in our ROSAT images was about $2 \times 10^{-14}\,\mathrm{erg\,cm^{-2}\,sec^{-1}}$ in the 0.1 - 2.4 keV band. Fits to the spectra of the stronger ROSAT sources with Raymond-Smith models gave typical spectral parameters of $kT \sim 2$ keV and $N_{\rm H} \sim 5 \times 10^{21}\,\mathrm{cm^{-2}}$. Using these values, the minimum detectable X-ray luminosity in our ROSAT images was about $5 \times 10^{29}\,\mathrm{erg\,sec^{-1}}$. A detailed comparison of the X-ray and optical data showed that we detected X-ray emission from nearly all optically selected stars earlier than about K5, many late K stars, but only a very small fraction of the M stars. This means that our current X-ray/optical sample is complete down to stellar masses of about $0.7\,M_\odot$.

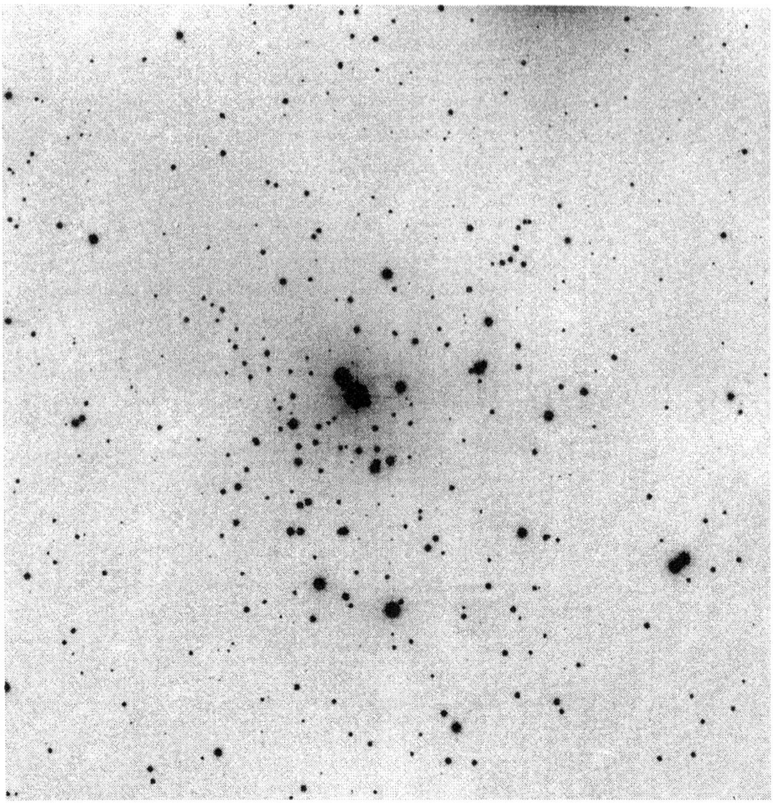

Fig. 34. Deep K-band image of IC 348, showing a 14 × 14 arcmin field.

Aims of the Proposed AXAF Observation. Our aim is to detect the *full* population of T Tauri stars in IC 348, *i.e.* down to $0.1\,M_\odot$. We can calculate the required X-ray sensitivity in the following way: According to the models of D'Antona & Mazzitelli [1], a $0.1\,M_\odot$ star with an age of 3 Myr has a bolometric luminosity of $\sim 0.03\,L_\odot$. If we assume a typical ratio $\log(L_X/L_{bol}) \sim -3$ for these stars, we can expect an X-ray luminosity of $L_X \sim 1 \times 10^{29}\,\mathrm{erg\,sec^{-1}}$. Since we also would like to detect stars which suffer somewhat stronger extinction or are slightly less active in X-rays, we aim at a minimum detectable X-ray luminosity of about $5 \times 10^{28}\,\mathrm{erg\,sec^{-1}}$ in the 0.1 - 2.4 keV band.

In order to reach our aims we need X-ray data which are one order of magnitude more sensitive than our previous ROSAT data. Additionally, we need high spatial resolution in order to be able to resolve and properly identify the numerous faint X-ray sources in the center of the cluster. Only AXAF/ACIS can provide us with data of sufficient quality to reach these aims. As we show below, the capabilities of AXAF/ACIS perfectly match our needs. With the proposed observation we expect to identify at least 100, possibly more than

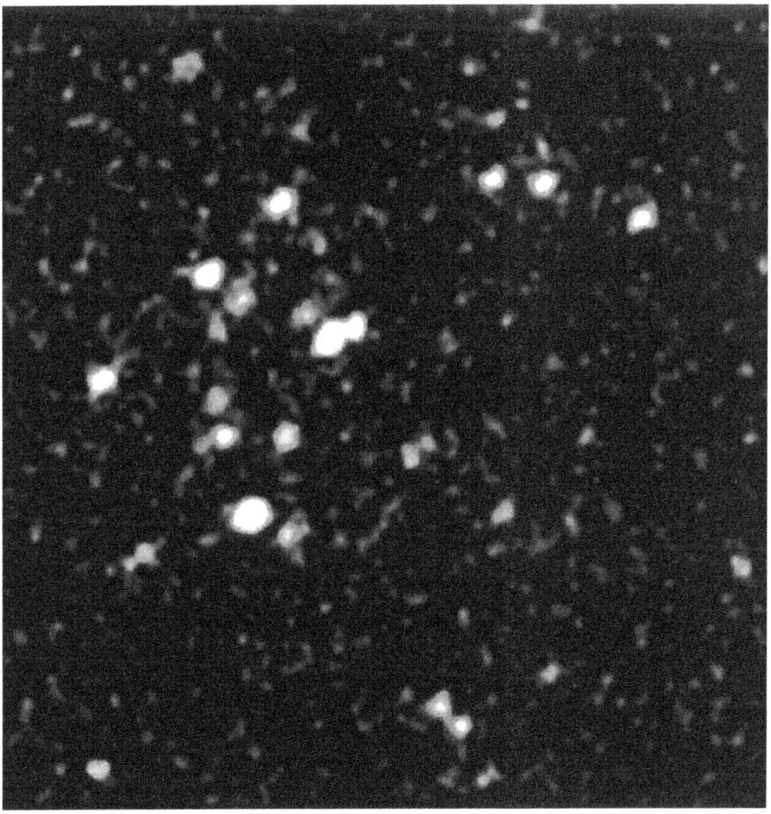

Fig. 35. Corresponding ROSAT HRI image of IC 348 (9.5×9.5 arcmin)

200 previously undetected faint X-ray sources as new cluster members. This X-ray selected sample will then serve as the basis for further investigations of the T Tauri star population in IC 348, which finally will yield the mass function and star formation history of the cluster. Additionally, the proposed observation will provide us with X-ray data on a large and homogeneous sample of T Tauri stars. For example, the data will be very well suited to study the temporal variability of the X-ray sources. About half of the ROSAT sources in IC 348 displayed significant variability and flares. The classical T Tauri star LHα 92 even showed the most energetic X-ray flare ever observed on a T Tauri star during our ROSAT observation [9]. Due to the large number ($\geq$ 200) of X-ray emitting T Tauri stars we expect to see at least a dozen large flares during our exposure. It would be very interesting if we could detect further X-ray flares on classical T Tauri stars; the analysis of the flare parameters could help to decide whether their X-ray activity can be fully explained by scaled up solar-type dynamo activity or whether non-solar

magnetic structures, which couple the star to its circumstellar disk, might possibly be involved (*cf.* [11]; [12]). Finally, we can expect to get X-ray spectra with more than 1000 counts for at least 20 sources in IC 348. The analysis of these spectra will provide us with important information about the coronal properties of the very active T Tauri stars (*cf.* [8]).

11.3 Technical Feasibility

In order to reach our aims, we have to reach the required sensitivity, to be able to make proper source identifications, and to cover the full cluster with homogeneous sensitivity. We now show that a 50 ksec AXAF/ACIS pointing can meet all these requirements:
The flux limit for such an observation will be $\sim 3\times10^{-15}\,\mathrm{erg\,cm^{-2}\,sec^{-1}}$ in the 0.2 - 10 keV band for a $5\,\sigma$ detection (AXAF Observatory Guide). Using the PIMMS simulator and assuming a Raymond-Smith thermal plasma spectrum with $kT = 2\,\mathrm{keV}$ and $N_\mathrm{H} = 5\times10^{21}\,\mathrm{cm^{-2}}$ (see above), we find that this corresponds to a flux of $\sim 1.5\times10^{-15}\,\mathrm{erg\,cm^{-2}\,sec^{-1}}$ in the 0.1 - 2.4 keV band. This flux transforms into a minimum detectable dereddened X-ray luminosity of $5\times10^{28}\,\mathrm{erg\,sec^{-1}}$ at the distance of IC 348, just as required. The projected star density in the central regions of our K-band image is some 50 stars per square-arcmin, *i.e.* the typical angular separation of the stars is about 5 arcsec. Thus, the fine spatial resolution of AXAF/ACIS of 1 - 2 arcsec will allow us to properly separate the X-ray sources and identify them with the stars in the optical and infrared images. Since the cluster covers an area of about 10×10 arcmin [4], the 17×17 arcmin ACIS field of view is perfectly suited to image IC 348.

11.4 Results

We have obtained a deep (50 ksec) X-ray image of the very young stellar cluster IC 348 with the Advanced CCD Imaging Spectrometer on AXAF (now called the CHANDRA X-Ray Observatory). In our image with a sensitivity limit of 1×10^{28} ergs sec^{-1} (more than 10 times deeper than our ROSAT images of IC 348), 215 X-ray sources are detected. While 115 of these sources can be identified with known cluster members, more than 50 X-ray sources are new, still unidentified cluster members. About 80 % of all known cluster members with masses between 0.15 and $2\,M_\odot$ are seen as X-ray sources in our high resolution image. We discover X-ray emission from four of 13 known brown dwarfs and from three of 12 brown dwarf candidates in IC 348. We also detect X-ray emission from two deeply embedded objects, presumably class I protostars, south of the cluster center.

Fig. 36. Deep CHANDRA X-ray image of the IC 348 cluster core (≈ 9 × 7 arcmin). This high spatial resolution X-ray image looks almost like an optical/infrared image. Comparison to Fig. 34 reveals a very similar pattern of stars, although some bright stars obviously are not detected as X-ray sources. (See: Th. Preibisch, H. Zinnecker: 'Deep CHANDRA X-Ray Observatory Imaging Study of the Very Young Stellar Cluster IC 348', AJ **122**, 866 (2001))

References

1. F. D'Antona, I. Mazzitelli: ApJS **90**, 467 (1994)
2. D. L. Harris, W. W. Morgan, N. G. Roman: ApJ **119**, 622 (1954)
3. G. H. Herbig: PASP **66**, 19 (1954)
4. G. H. Herbig: ApJ **497**, 736 (1998)
5. J. Krautter: 'The impact of ROSAT observations on our understanding of star forming regions'. In: *Cool stars; stellar systems; and the sun : 9 : Astronomical Society of the Pacific Conference Series, volume 109; Proceedings of the 9th Cambridge workshop; held 3-6 October 1995 in Florence; Italy*, ed. by R. Pallavicini, A. K. Dupree (Astronomical Society of the Pacific, ASP, San Francisco 1996) p. 395
6. E. A. Lada, C. J. Lada: AJ **109**, 1682 (1995)
7. Th. Montmerle: 'X-rays from young stellar objects: from T Tauri stars to protostars'. In: *Cool stars; stellar systems; and the sun : 9 : Astronomical Society of the Pacific Conference Series, volume 109; Proceedings of the 9th Cambridge workshop; held 3-6 October 1995 in Florence; Italy*, ed. by R. Pallavicini,

A. K. Dupree (Astronomical Society of the Pacific, ASP, San Francisco 1996) p. 405
8. Th. Preibisch: A&A **320**, 525 (1997)
9. Th. Preibisch, H. Zinnecker, J. H. M. M. Schmitt: A&A **279**, L33 (1993)
10. Th. Preibisch, H. Zinnecker, G. H. Herbig: A&A **310**, 456 (1996)
11. F. H. Shu, H. Shang, A. E. Glassgold, T. Lee: Science **277**, 1475 (1997)
12. F. M. Walter, P. B. Byrne: 'A New Non-Solar Paradigm for Active Stars'. In: *The Tenth Cambridge Workshop on Cool Stars, Stellar Systems and the Sun*, ed. by R. A. Donahue, J. A. Bookbinder; ASP Conference Series **154**, 1458 (1998)

12 A Study of the Low-Mass Stellar Population in the Galactic Starburst Region NGC 3603

12.1 Abstract

NGC 3603 is the most vigorous, visible Galactic starburst region with the highest concentration of very massive stars known in the Galaxy. While its population of high and intermediate mass stars has been previously characterized, the content of low mass stars is fairly unknown. Our main aim here is to investigate the relation between high and low mass star formation and — in the general context of starbursts — whether there are lots of low-mass stars in starburst regions or not. Its relative proximity and young age make NGC 3603 the ideal target for a study in the infrared JHK bands. Under very good seeing conditions, the excellent sensitivity *and* spatial resolution of ISAAC on the VLT would enable us to determine the number *and* spatial distribution of the stellar populations well below $0.08 M_\odot$ (K $\approx$ 20) — deeper than ever before in a giant HII region.

12.2 Description of the Proposed Programme

Scientific Rationale. The main goal of this project is to better understand the processes that lead to the stellar mass spectrum in starburst regions. Do low-mass stars really dominate the mass of starburst clusters as inferred from the extrapolation of their IMF slopes? Do most stellar members of a cluster form simultaneously, or do low- and high-mass stars form independently in time and space via different physical processes? Is there a universal IMF, and if so what does it look like?

Apparently low mass stars form at a relatively low, almost constant rate in molecular clouds, as observed, *e.g.*, in IC 348 [11]. This slow mode of star formation, probably regulated by ambipolar diffusion [18], may be quite different from the process that leads to clusters of very massive stars: luminous, young star clusters like the Trapezium cluster, R 136, or NGC 3603 appear to have formed within less than 1 Myr, a timescale that is too short to be solely accounted for by ambipolar diffusion. If there are two modes of star formation, they may show different characteristics not only in the duration of formation but also in the stellar mass spectrum. The possibility that violent star formation initially favours massive stars was first suggested by [17]. Variations in the ^{16}O abundance in the Galaxy suggest a truncated ($\leq 2\,M_\odot$) IMF in the spiral arms where enhanced star formation takes place, but not in the interarm regions [6]. Observational results from *integrated* luminosities and line fluxes in distant starbursts like M 82 [16] or M 83 [9] indicate a lower mass cutoff of a few solar masses. Also, theoretical investigations of Silk [19]; [20] and Larson [12] show that the Jeans mass increases when newly born, massive stars heat the surrounding gas, suppressing the formation of low-mass stars in their environment. On the other hand, recent studies of nearby young clusters such as the Trapezium cluster have detected a sub-solar mass population ($\leq 0.2\,M_\odot$, [7], see also the overview article by Zinnecker, McCaughrean, & Wilking in [25]). To resolve this controversy, precise and complete photometric information on individual, faint stellar sources in an extreme nearby starburst is required.

Immediate Objective. We propose VLT/ISAAC observations to detect the number *and* spatial distribution of the low-mass stars in the galactic starburst template NGC 3603. This young cluster is the ideal target for our study: it is the most massive, optically visible HII-region in our Galaxy, located in the Carina spiral arm at a distance of only ~ 7 kpc. Within a diameter of 1 parsec (30 arcsec) the mass in OB-stars is $\approx 1000\,M_\odot$ with a peak mass density in the core of $10^5\,M_\odot\,\mathrm{pc}^{-3}$, making it the densest concentration of massive stars known in the Galaxy [4] – a real starburst!

NGC 3603 has been a favorite target in numerous high spatial resolution observations, using ground based speckle interferometry [8] as well as the WFPC [14] and FOS [4]. These observations, however, yielded complete results only for OB-stars.

NGC 3603 is a galactic analog to the prototype starburst cluster R 136 in the LMC. Indeed NGC 3603 and R 136 (30 Dor) are similar in many respects, such as age and central density; as in R 136, several molecular clouds around NGC 3603 show ongoing star formation [13]. NGC 3603 in fact looks like a clone of the core of the 30 Doradus region, R 136, but without its extended, surrounding halo [14]. Many specific properties justify an independent and detailed study, *e.g.*, the central WR stars in NGC 3603 are on average about one magnitude brighter than those in R 136, maybe due to different initial starburst conditions. Thus comparing the results on the mass function from at least one extragalactic (R 136) and one galactic region (NGC 3603) is essential to establish a general picture of the stellar content in starbursts (analysis of our R 136 data is in progress).

From the observational point of view, the most exciting difference is the relative proximity; the fact that NGC 3603 is about 7 times closer than R 136 not only enables us to detect much fainter, less-massive stars but also re-

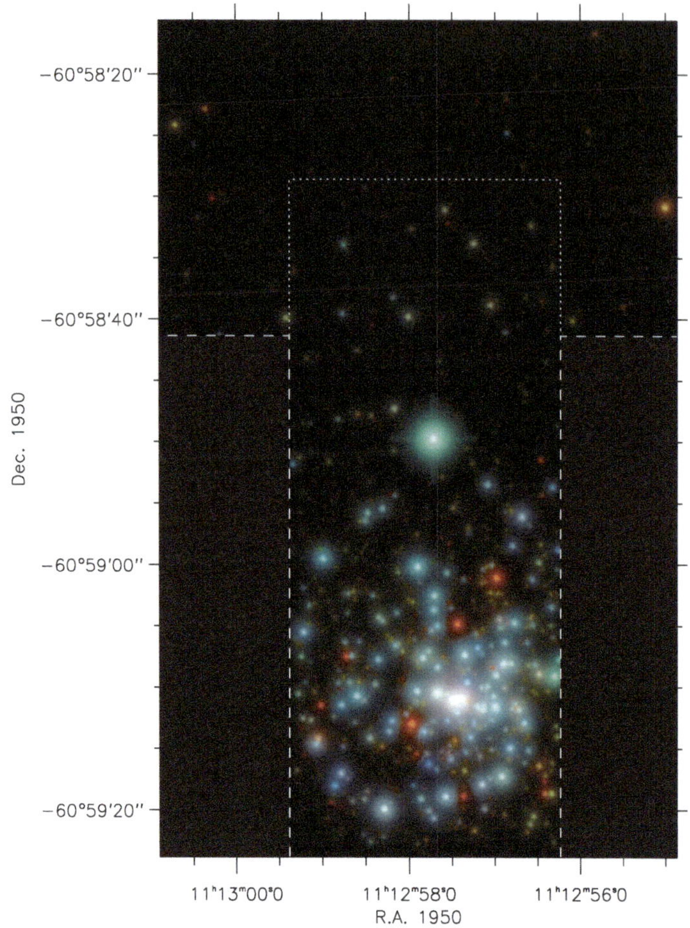

Fig. 37. Stars in NGC 3603 detected in Eisenhauer *et al.'s* [5] near-infrared ADONIS observations. The faintest stars detected here are about 19th magnitude.

duces the significant problem of crowding and photometric errors. Recently, Eisenhauer *et al.* [5] reported on a study in J-, H- and K-bands of about 900 stars in the central region of NGC 3603, based on adaptive optics observations with ESO's system ADONIS/SHARP2. Eisenhauer *et al.'s* data indicate no turnover or truncation of the IMF slope down to at least $1\,M_\odot$, where the data sample suffers from incompleteness. The observations are limited to about 19th magnitude in JHK and spatial resolutions of 0.18 arcsec, 0.30 arcsec, and 0.31 arcsec in J, H, and K-bands, respectively. As can be easily seen in Fig. 37, the observations are limited in sensitivity and by blending: the typical PSF shape in adaptive optics observations contains a large percentage of the source flux in the extended seeing disk, which is also plagued by variable, residual speckles, making the detection of very faint sources in the vicinity of bright stars almost impossible (somewhat like HST before COSTAR and WFPC2).

If we extrapolate Eisenhauer's IMF slope of $\Gamma = -0.7$ to $0.08\,M_\odot$ we would expect approximately 3300 stars between $1\,M_\odot$ and $0.08\,M_\odot$ within the same region of 2190 arcsec^2 — *i.e.*, about 2 such low-mass objects per square arcsecond on average. While the immediate cluster core is certainly very dense, we do not expect to be limited by crowding outside one core radius of NGC 3603. Our proposed observations will be comparable in spatial resolution with the central core in Eisenhauer *et al.'s* adaptive optics observations, but without the surrounding seeing halo, thus yielding much higher point source sensitivity.

The advantage of performing the proposed observations at infrared wavelengths is twofold: First, the dust extinction toward NGC 3603 at K-band is only 0.5 magnitudes with respect to 4.6 in V-band. Since most directly observable star forming regions are located in the Galactic plane one typically gains one to two magnitudes in respect to optical observations. Secondly, because of the young cluster age, stars less massive than $2\,M_\odot$ are still in their pre-main-sequence (PMS) phase, probably surrounded by disks that emit infrared excess. Kenyon & Hartmann [10] have measured colors of low-mass stars with circumstellar disks in Taurus-Auriga: $(V-K) \approx 4^m - 7^m$. Thus the total gain for these stars in the NIR may be in the order of $\sim$5 - 8 magnitudes, relative to the visible.

We propose to observe NGC 3603 in the near-infrared J, H, and K bands, that provide a unique tool to determine extinction and infrared excess emission for each individual star [21] — a possibility not available at visible wavelengths. Three wavebands (*i.e.*, two colors) yield a very good estimate of the stellar luminosities and provide sufficient information for precise mass determinations, even for PMS stars (one color only would not be sufficient to identify field stars or disentangle local extinction from stellar luminosity without additional assumptions).

12.3 Required Observing Time

We derive the absolute magnitudes of a $0.08\,M_\odot$ star following Testi, Palla, & Natta [22]:
$M_K = 4.725 - 2.5\log(L_\star/L_\odot) - \mathrm{BC}_V(T_{eff}) - (V-K)(T_{eff}) \approx 4.8$, based on the PMS evolutionary tracks of D'Antona & Mazzitelli [1]; the bolometric correction and colors are taken from Testi, Palla, & Natta [22]. We further assume an effective temperature of 2950 K and an age of about 0.5 Myr for the subsolar mass population (see Fig. 38). For a distance modulus of m − M = 14.3 and global reddening of $A_V = 4.6$ [5], we calculate apparent magnitudes of $m_J = 21.2$, $m_H = 20.5$, and $m_K = 19.6$. For these values, ISAAC's exposure time calculator yields $t_J = 900sec$, $t_H = 2500sec$, and $t_K = 1700\,sec$ on-source for a S/N = 50 point source detection at airmass 1.3

and seeing of 0.4 arcsec. This is based on detector integration times of 0.5 sec to keep saturation by the most luminous stars in the cluster to a minimum. We assume a signal-to-noise ratio of 50 for the point source sensitivity to be a lower limit to include the effect of reduced sensitivity by blending.

The minimum integration time totals ~5000 sec; given some uncertainty in the PMS model calculations we request 2 hours of on-source integration time. (This does not include integration time for sky-background determination which is best done by mapping the sky around NGC 3603 and calculating the average sky flux using the median technique). ISAAC's field of view of 150 × 150 arcsec is sufficient to cover the region of interest in one pointing; no mapping is required.

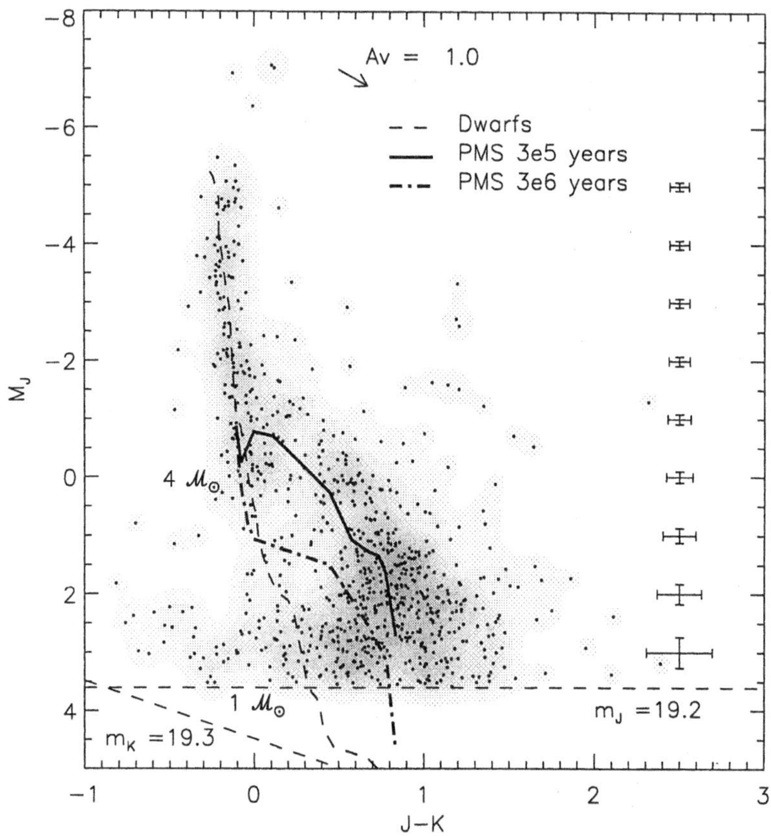

Fig. 38. Color magnitude diagram of 820 stars within the central region of NGC 3603. The diagram includes theoretical pre-main-sequence isochrones and the main sequence (from [5]).

Fig. 39. Near-infrared image of NGC 3603 obtained with ISAAC at the VLT. FOV is 3.4 arcmin × 3.4 arcmin (6.2 × 6.2 parsec^2). North is up, East to the left. The insert to the lower right is a blow up of the central parsec^2 (Brandl et al. 1999).

12.4 Results

We have used the near-infrared camera ISAAC on VLT/UT1 to observe NGC 3603, the most massive visible H II region known in the Galaxy, in very good seeing (0.4 arcsec) (see Fig. 39). Our J, H, K observations are the most sensitive, high spatial resolution observations made to date of this dense starburst region, allowing us to resolve its low-mass stellar population down to $0.1\,M_{\odot}$. The overall age of the pre-main sequence stars in the core region of NGC 3603 derived from isochrone fitting is of the order of 1 Myr. The near-infrared luminosity functions show that the cluster is indeed populated in low-mass stars at least down to $0.1\,M_{\odot}$. Our observations clearly show that sub-solar mass stars do form in massive starbursts.

Acknowledgments. This proposal was written by B. Brandl (PI), with HZ as one of the co-investigators. The results are published in Brandl et al. (1999) In: A&A **352**, L69. See also: B. Brandl, H. Zinnecker, W. Brandner: 'The IMF in Starbursts'. In: *Star Formation 1999, Proceedings of Star Formation 1999, held in Nagoya, Japan, June 21 - 25, 1999*, ed. by T. Nakamoto, pp.341; and B. Brandl et al.: *IAU Symposium 207, Extragalactic Star Clusters*, in press.

References

1. F. D'Antona, I. Mazzitelli: ApJS **90**, 467 (1994)
2. B. Brandl, B. J. Sams, F. Bertoldi et al.: ApJ **466**, 254 (1996)
3. W. Brandner, E. K. Grebel, Y.-H. Chu, K. Weis: ApJ **475**, 45 (1997)
4. L. Drissen, A. F. J. Moffat, N. R. Walborn, M. M. Shara: AJ **110**, 2235 (1995)
5. F. Eisenhauer, A. Quirrenbach, H. Zinnecker, R. Genzel: ApJ **498**, 278 (1998)
6. R. Güsten, P. G. Mezger: Vistas in Astronomy **29**, 159 (1983)
7. L. A. Hillenbrand: AJ **113**, 1733 (1997)
8. K.-H. Hofmann, W. Seggewiss, G. Weigelt: A&A **300**, 403 (1995)
9. E. B. Jensen, R. J. Talbot Jr., R. J. Dufour: ApJ **243**, 716 (1981)
10. S. J. Kenyon, L. Hartmann: ApJS **101**, 117 (1995)
11. E. A. Lada, Ch. J. Lada: AJ **109**, 1682 (1995)
12. R. B. Larson: MNRAS **214**, 379 (1985)
13. J. Melnick, M. Tapia, R. Terlevich: A&A **213**, 89 (1989)
14. A. F. J. Moffat, L. Drissen, M. M. Shara: ApJ **436**, 183 (1994)
15. A. F. J. Moffat: RMxAC **6**, 108 (1997)
16. G. H. Rieke, M. J. Lebofsky, R. I. Thompson et al.: ApJ **238**, 24 (1980)
17. M. Schwarzschild, L. Spitzer: Obs **73**, 77 (1953)
18. F. H. Shu, F. C. Adams, S. Lizano: ARA&A **25**, 23 (1987)
19. J. Silk: ApJ **214**, 718 (1977)
20. J. Silk: ApJ **438**, 41 (1995)
21. K. M. Strom, S. E. Strom, K. M. Merrill: ApJ **412**, 233 (1993)
22. L. Testi, F. Palla, A. Natta: A&AS, **133**, 81 (1998)
23. D. J. Wilner, Ch. J. Lada: AJ **102**, 1050 (1991)
24. H. Zinnecker, A. Moneti: 'NICMOS/HST H-band (F 160W) Images of the 30 Dor Starburst Cluster'. In: *NICMOS and the VLT, A New Era of High Resolution Near Infrared Imaging and Spectroscopy, Workshop in Pula, Italy, May 26-27, 1998*, ed. by W. Freudling, R. Hook, ESO Conference and Workshop Procedings, **55**, 136 (1998)
25. H. Zinnecker, M. J. McCaughrean, B. A. Wilking: 'The initial stellar population'. In: *Protostars and Planets III* (Univ. of Arizona Press), eds. E.G. Levy and J. Lunine, p. 429 (1993)

13 The Low-Mass Pre-MS Stellar Content of the 30 Dor Starburst Cluster

13.1 Abstract

We propose to take diffraction limited near-infrared H-band (1.6 μm) NICMOS images of the young NGC 2070 cluster (age 3.5 Myr) in the 30 Dor giant HII region in the LMC. The aim is to search for the low-mass ($M < 2\,M_\odot$) low-luminosity, red pre-Main Sequence stellar population and to establish the H-band infrared luminosity function. With the NICMOS we can now determine whether the IMF in this prototypical extragalactic starburst cluster is deficient in subsolar low-mass stars or not. The best ground-based data can sample only $M > 2\,M_\odot$. In principle, NICMOS in the H-band (F160W) is sensitive enough to reach a magnitude of ~ 23.5 in a relatively short integration time, which indeed corresponds to the fantastic possibility to detect young stellar objects with masses near the hydrogen burning limit ($M = 0.1\,M_\odot$) according to pre-Main Sequence evolutionary models. Even if we could reach only $H = 22.5$ (*i.e.* $M = 0.4\,M_\odot$), the observations proposed here would still go a long way in directly answering, by star counts, whether the IMF in starburst galaxies is low-mass deficient or not, with all the corresponding far-reaching implications. The observations would also tell us whether the 30 Dor cluster can be regarded as a prototype young globular cluster. This possibility would be ruled out, if we found NGC 2070 to be low-mass deficient, because old globular clusters do have a rich population of low-mass stars.

13.2 Scientific Background

The 30 Dor Cluster. 30 Dor in the LMC is the most luminous giant HII region in the Local Group [6]; [7]; see Fig. 40. It has been called "the starburst Rosetta Stone" [17]. It is powered by hundreds of massive stars in the core of the NGC 2070 young star cluster (hereafter "the 30 Dor cluster"). The core is known as R 136 and is only a few parsec in size, while the diameter of the full cluster is 40 pc. The age of the 30 Dor cluster is around 3.5 Myr (judged from the presence of Wolf-Rayet stars and the absence of red supergiants); the age spread is small, less than 1 Myr, so to first order the cluster has a coeval high mass stellar population ($M > 15\,M_\odot$). Recent WFPC2 HST data [5] have established the presence of an intermediate mass population ($3 - 15\,M_\odot$) and a slope of the IMF close to the Salpeter slope. Furthermore, Brandl *et al.* [1] from ground-based H and K band adaptive optics imaging, found a somewhat steeper and spatially variable slope. They probe the IMF down to $M \sim 2\,M_\odot$ (H and K near 20th mag). Nothing is known about the low-mass population ($M < 2\,M_\odot$). With NICMOS on HST it is now possible to determine the low-mass stellar content. Thus the present application is for NICMOS/HST time to establish if there is or if there is not a major low-mass population ($0.1 - 2\,M_\odot$) in the 30 Dor cluster.

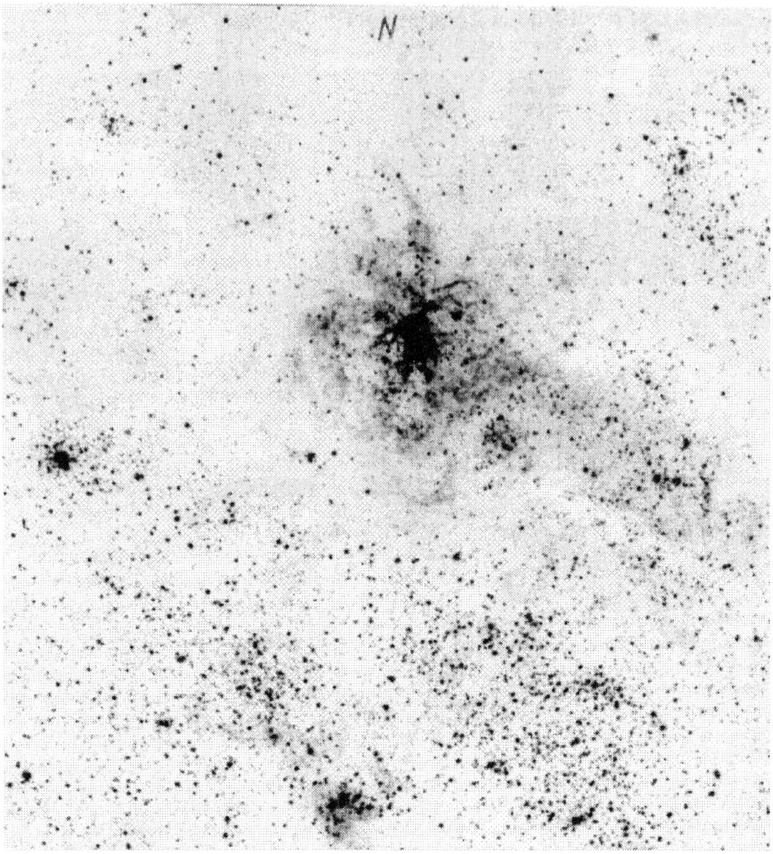

Fig. 40. The 30 Doradus HII region, also known as the "Tarantula Nebula", is shown near the center of this image, taken from the Digital Sky Survey. This giant HII region is powered by the massive cluster NGC 2070. The image size is ~40 arcmin × 40 arcmin corresponding to 600 pc × 600 pc at the distance of the LMC. 20 arcmin to the east (left) lies the twin cluster NGC 2100 which is slightly older (10 Myr) and has already lost its HII region nebulosity.

Significance to Astronomy. Starburst clusters appear to be the primary sites (unit cells) of star formation in starburst galaxies, including interacting/colliding galaxies such as the Antennae or The Cartwheel. Recent HST images [20]; [2] have revealed hundreds of giant young star clusters in these and other starburst galaxies (*e.g.* M 82 [11]). If starburst clusters are the basic building blocks of certain star forming galaxies, their stellar content (IMF) will affect much of the observed chemical and photometric evolution of galaxies, both at the present epoch and perhaps even more so in the high-redshift past [3]. Thus the question whether the IMF is top-heavy in starburst galaxies (as has been frequently claimed, *e.g.* Rieke *et al.* [15] from indirect observations or Silk [16] on theoretical grounds) is essentially the question of whether the IMF is deficient in low-mass stars in starburst clusters. Note that a system with a top-heavy IMF will fade away faster than one with a standard IMF and may eventually disappear, while producing more heavy elements and more stellar remnants. Thus, provided that the 30 Dor cluster is a prototypical starburst cluster, it is clear that determining its low-mass stellar content is of far-reaching importance.

Another side of the same coin is the claim that starburst clusters represent young globular clusters. Indeed there is some resemblance [10]; [8]. However, for this to be true, starburst clusters should contain not only luminous high-mass stars but also low-mass pre-Main Sequence stars in substantial numbers. Indeed present-day old globular clusters are full of low mass stars (*e.g.* the HST images of NGC 6397 [13]). This begs the long-standing question of whether the 30 Dor cluster can be regarded as a prototype young globular cluster, albeit a small one. Many people have speculated that it might be, but there has always been the problem that low-mass stars have not yet been directly observed in 30 Dor. Therefore there is not yet any direct evidence to support this hypothesis. The simple observation of *any* significant number of low-mass stars in the 30 Dor cluster would greatly strengthen the case for regarding it as a prototype young globular cluster. With NICMOS on HST such observations can now be undertaken. These observations will also reveal the spatial distribution of stars of different mass in the cluster and will allow us to check whether the core radii for stars of different mass are different, hence whether some mass segregation has already occurred in this very young stellar system (*cf.* [1]).

13.3 Previous Work

Initially, high spatial resolution HST work [18]; [9] and ground-based speckle studies [14] of the 30 Dor cluster were limited to the most massive stars only. Weigelt and Baier [20] were the first to resolve the massive stars in R 136. Brandl *et al.* [1] discuss the comparison of their near-infrared adaptive optics data with the optical WFPC2 data of Hunter *et al.* [5]. With a few exceptions, all stars brighter than V = 16 were detected in the near-infrared. Almost 60 sources were found in the near-infrared which were not detected in the HST UVI bands. These may be the bright tail of pre-Main Sequence objects which we hope to discover in the thousands. Brandl *et al.* [1] provide evidence that the core radius grows with decreasing stellar mass (0.5 arcsec or 0.12 pc for $M \geq 15\,M_{\odot}$ and 1.0 arcsec or 0.25 pc for $M \geq 4\,M_{\odot}$), consistent with the slope of the IMF changing with radial distance from the cluster center. They speculate from timescale arguments that dynamical mass segregation has already occurred for the more massive stars and that this evolutionary effect might explain why the IMF is shallower in the inner part. On the other hand, Hunter *et al.* [5] see no evidence for mass segregation beyond 0.5 pc in their data. Clearly future observations must clarify this issue. Both studies agree, however, that the IMF is not truncated down to their respective completeness limits. This may suggest that the IMF continues smoothly to lower masses. The complication brought along with studies of masses below $2\,M_{\odot}$ is the need to understand pre-Main Sequence evolution. Pre-MS luminosity functions of coeval model star clusters (for ages of 1, 2, 3 Myr) have been constructed [21], and time-dependent luminosity-mass relations are at hand. Thus infrared J, H, or K luminosity functions can be converted to the corresponding mass function - and this is what we intend to do in the case of the largely coeval 30 Dor cluster.

13.4 Summary of the Scientific Goals and Method

The present proposal is for deep NICMOS observations to address/solve two simple but fundamental questions:

- is there a low-mass ($M < 1\,M_{\odot}$) stellar population in the 30 Dor starburst cluster or is the IMF deficient in low mass stars?
- can the 30 Dor cluster be regarded as a young globular cluster?

The method to be used is star counts at $1.6\,\mu$m, going to deeper and deeper magnitude bins (as deep as crowding allows) and witnessing whether the number of stars keeps increasing or not. The choice of the wavelength is crucial to the success of the proposed observations, because only in the F160W filter can we reach faint enough magnitudes that we can probe the low mass stellar content effectively (ideally $H = 24$ or $0.1\,M_{\odot}$, but even $H = 22.5$ or $0.4\,M_{\odot}$ would be an enormous progress).

13.5 The Need for NICMOS/HST

First of all, the need for infrared (NICMOS) observations arises from the fact that we are looking for low-mass stars which are cool objects ($V - K = 3.5$, $V - H = 3.3$ for a $0.5\,M_{\odot}$ star).

H-band observations are favoured over K-band observations due to the much lower background for NICMOS on HST at H than at K and also due to the higher spatial resolution for diffraction-limited imaging at H (~ 0.15 arcsec) compared to K (~ 0.25 arcsec). At the same time, observing in the near-infrared the patchy visual extinction in or over the face of the cluster ($A_V < 1.5$ mag) is much reduced ($A_H < 0.3$ mag). Taken together (red objects, extinction) the H-band observations have a 4.5 mag advantage over the visual observations in this case, even for ground-based observations.

The need for HST observations comes primarily from the fact that we want to search for low-mass objects, too faint to be detected from the ground, even with high-resolution adaptive optics techniques which exhaust their potential around $K = 20$ or $H = 20$ (see [1]). There is an important reason for this:

Although the point spread function of an adaptive optics system is quite spiky, the wings typically contain more than 50 % of the source flux spread out over a few arcsec. This poses no problem for resolving sources of similar brightness, but severe problems for the detection of faint stars in the vicinity of brighter sources (*i.e.* almost everywhere). In contrast, 93 % of the total flux is concentrated within only 0.3 arcsec of the well-defined point spread function of the HST.

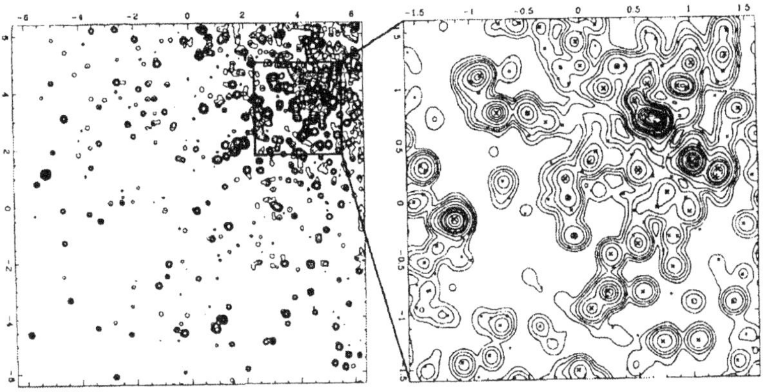

Fig. 41. 190 min K band adaptive optics exposure of R 136 from Brandl *et al.* [1]. The field of view is 12.8 × 12.8 arcsec corresponding to 3.2 × 3.2 pc. North is up, east to the left. The spatial resolution after deconvolution is 0.15 arcsec FWHM; scaling is logarithmic. The faintest stars seen have 20th magnitude in K. This is very similar what NICMOS camera 1 on HST will see, albeit at H, after a much shorter integration.

High spatial resolution is needed because of the crowding problem in this dense cluster, especially for low-mass stars. An impression of the expected crowding can be gained from the 30 Dor adaptive optics image at K taken at the ESO 3.6 m telescope (Fig. 41) which has the same spatial resolution as we would get at H with the 2.4 m HST; the limiting magnitude of the ESO image is K ∼ 20, while our proposed HST observations would go a factor of 10 or more deeper, as we want to search for fainter lower mass stars. Clearly, even with the spatial resolution that HST provides, crowding remains an issue. The reliability of the so-called incompleteness correction, however, increases significantly with the spatial resolution and would be a hopeless undertaking with seeing limited observations.

13.6 Description of the Observations

The observations will be carried out with NICMOS in the F160W filter, using both NIC1 (0.043 arcsec/pixel) in the innermost 20 × 20 arcsec heavily crowded region and NIC2 (0.075 arcsec/pixel) in the outer less crowded regions. The inner region will be covered by a 2×2 mosaic of NIC1 frames, plus a central NIC1 frame centred on R 136a. The outer region will be covered by a 3×3 mosaic of NIC2 frames (covering roughly the central square arcmin of the cluster, corresponding to about 7.5×7.5 pc). In addition, we will add symmetrically two linear extensions of three frames each (see figure in [22]) to fully cover one diameter (160 arcsec or 40 pc) of the cluster. The observations will be performed in MULTIACCUM mode, to be able to deal with the saturation of bright objects in long exposure frames. In addition, 6 sky frames (3 with NIC1 and 3 with NIC2) will be taken some 5 - 10 arcmin north of the cluster and will be used both for sky subtraction and for determining the density of foreground/background sources.

At each position, we plan to take 4 dithered images of 10 - 12 min duration each (the exact time will be determined in Phase II, and will be adjusted for the 4 images and overheads to fill the orbit) which will be used 1) to sample the PSF more finely than with a single image, and 2) to reject cosmic ray events. Using the exposure time estimators we determined that on single images we would reach $H \approx 23.5$ mag and $J \approx 24.5$ mag with $\sim 4\,\sigma$ on an isolated source. In practice source crowding will be very important, especially in the cluster center, where it will undoubtedly be the limiting factor. Additionally, our final S/N ratio will also be determined by the amount of extended emission from the HII region. This we estimated from our own ground-based images: it will be $\sim 24-25$ mag/pixel in NIC2/F160W on average, though brighter in a few positions. We estimate that with 4 images per position we will reach close to $H = 23.5$ mag over most of the field covered.

We also plan to use NIC1/F110W filter while observing with NIC2/F160W, and vice-versa, which will give us further precious information. This will, in the end, give us a significant amount of J photometry in addition to the H data which is the main goal of our observations.

13.7 Results

We present the first infrared luminosity function in the 30 Doradus star cluster obtained with the HST NIC1 camera (0.043 arcsec / pixel) in the F160W (H-band) filter. Despite diffraction limited resolution (0.15 arcsec), crowding and blending is so severe that the cluster centre R136 cannot be studied. Instead a neighbouring NIC1 field (about 15 arcsec away from the centre), for which an empirical point spread function could be obtained, was analysed. We obtained photometry for some 750 stars in the range $H = 14$ to 22 mag. The luminosity function continues rising up to $H = 21$ mag ($1\,M_\odot$ at age 2 Myr) after which incompleteness corrections become dramatic (exceeding 50 %). Contrary to recent optical studies based on a V vs. V-I diagram, which suffer from extinction problems, we do not infer a flattening or turnover of the implied IMF slope near $2\,M_\odot$ (Fig. 43).

Fig. 42. H-band (F 160W) NICMOS/HST image of the core of NGC 2070. The image is a 3 × 3 mosaic of NIC2 frames, covering a total field of 1 × 1 arcmin (15 × 15 pc at the distance of the Large Magellanic Cloud). The spatial resolution is 0.15 arcsec (HST diffraction limit at 1.6 μm). The field of interest is marked by the box.

References

1. B. Brandl, B. Sams, F. Bertoldi, A. Eckart, R. Genzel et al.: ApJ **466**, 254 (1996)
2. K. D. Borne et al.: Haro Starburst Conference, Puebla/Mexico (1996)
3. S. Charlot, F. Ferrari, G. J. Mathews, J. Silk: ApJ **419**, L57 (1993)
4. A. Forestini: A&A **285**, 473 (1994)
5. D. Hunter, E. J. Shaya, J. A. Holtzman et al.: ApJ, **448**, 179 (1995)
6. R. C. Kennicutt: ApJ, **287**, 116 (1984)
7. R. C. Kennicutt, Y.- H. Chu: In: *Violent Star Formation: from 30 Doradus to QSOs*, ed. by G. Tenorio-Tagle (Cambridge University Press, Cambridge 1994) p. 1
8. R. B. Larson: 'Present-Day Cluster Formation In: *The Globular Cluster - Galaxy Connection. Astronomical Society of the Pacific Conference Series, Volume 48,*

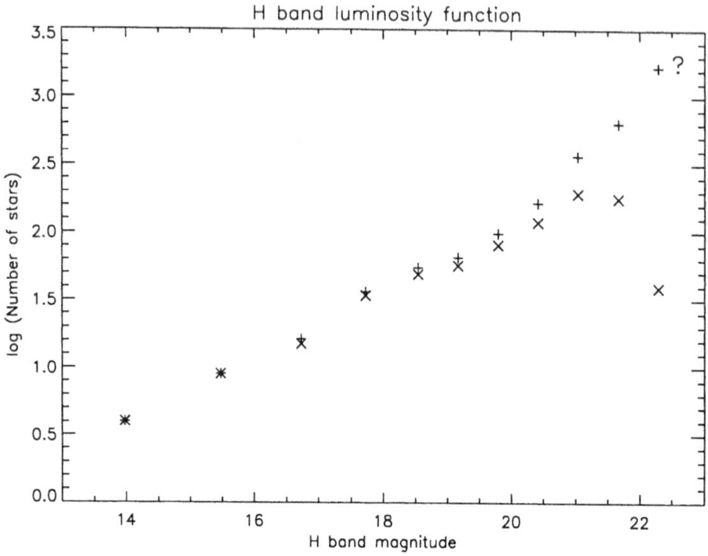

Fig. 43. Infrared H-band stellar luminosity function of the outskirts of the 30 Dor starburst cluster based on photometry of objects in the square area as shown in the HST/NICMOS image in Fig. 42. The crosses are the uncorrected data, the plus signs are the data corrected for incompleteness. Note that the luminosity function is still rising at the faint end (from H. Zinnecker, M. Andersen et al.: *IAU Symposium 207*, in press; see also: H. Zinnecker, B. Brandl, W. Brandner et al.: '30 Doradus: The Low-Mass Stars'. In: *New Views of the Magellanic Clouds, IAU Symposium 190*, ed. by Y.-H. Chu, N. Suntzeff, J. Hesser, D. Bohlender, pp.222

Proceedings of the 11th Santa Cruz Summer Workshop in Astronomy and Astrophysics, held July 19-29, 1992, at the University of California, Santa Cruz, ed. by G. H. Smith & J. P. Brodie (Astronomical Society of the Pacific, ASP, San Francisco 1993) p. 675

9. E. M. Malumuth, S. R. Heap: AJ, **107**, 1054 (1994)
10. G. Meylan: 'The Properties of 30 Doradus'. In: *The Globular Cluster - Galaxy Connection. Astronomical Society of the Pacific Conference Series, Volume 48, Proceedings of the 11th Santa Cruz Summer Workshop in Astronomy and Astrophysics, held July 19-29, 1992, at the University of California, Santa Cruz*, ed. by G. H. Smith & J. P. Brodie (Astronomical Society of the Pacific, ASP, San Francisco 1993) p. 588
11. R. W. O'Connell, J. S. Gallagher, D. A. Hunter, W. N. Colley: ApJ, **446**, L1 (1995)
12. F. Palla, S. W. Stahler: ApJ, **418**, 414 (1993)
13. F. Paresce, G. de Marchi, M. Romaniello: ApJ, **440**, 216 (1995)
14. E. Pehlemann, K.-H. Hofmann, G. Weigelt: A&A, **256**, 701 (1992)
15. G. H. Rieke, K. Loken, M. J. Rieke, P. Tamblyn: ApJ, **412**, 99 (1993)
16. J. Silk: ApJ, **438**, L41 (1995)

17. N. R. Walborn: In: *Massive Stars in Starbursts*, ed. by C. Leitherer et al. (Cambridge University Press, Cambridge 1991) p. 145
18. G. Weigelt, R. Albrecht, C. Barbieri et al.: ApJ, **378**, L21 (1991)
19. G. Weigelt, G. Baier: A&A, **150**, 18 (1985)
20. B. C. Whitmore, F. Schweizer: AJ, **109**, 960 (1995)
21. H. Zinnecker, M. J. McCaughrean, B. A. Wilking: 'The initial stellar population'. In: *Protostars and Planets III* (Univ. of Arizona Press), eds. E.G. Levy and J. Lunine, p. 429 (1993)
22. H. Zinnecker, A. Moneti: 'NICMOS/HST H-band (F160W) Images of the 30 DOR Starburst Cluster'. In: *NICMOS and the VLT: A New Era of High Resolution Near Infrared Imaging and Spectroscopy, Pula, Sardinia, Italy, June 26-27th 1998*, ed. by W. Freudling & R. Hook (ESO Conference and Workshop Proceedings 55, 1998) p. 136

V. Appendix: The Suite of Large Ground-Based Telescopes

Overview of the all large telescopes currently available or under construction: it is seen that there are more 8 m class telescopes these days than 4 m class telescopes! (note for students: the golden age of astronomy is **now**)

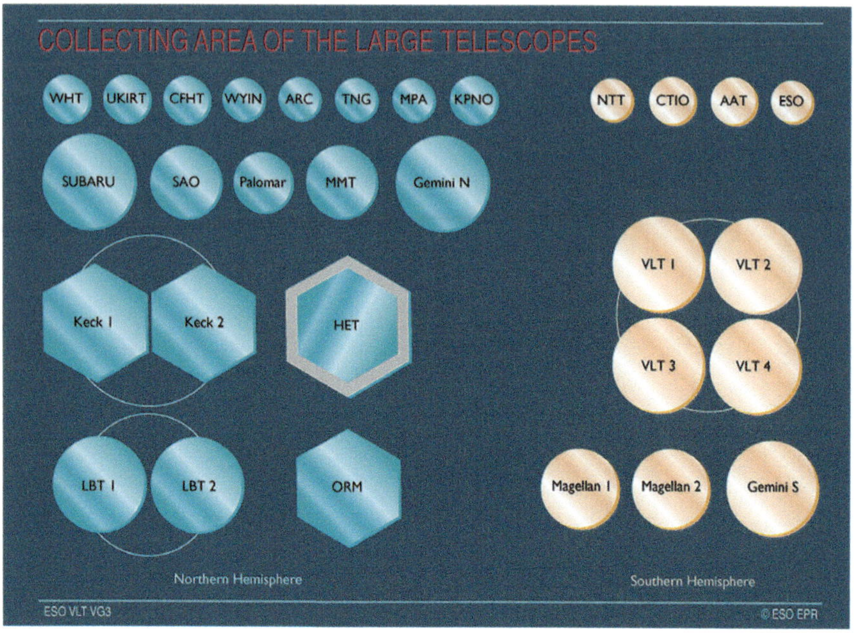

Fig. 44. The collecting area of existing and future, large telescopes in the Northern Hemisphere (left) and in the Southern Hemisphere (right). Keck I and II, HET and ORM have segmented mirrors; the others have monolithic mirrors (courtesy ESO/EPR).

Artist's view of the Atacama Large Millimeter Array - a mm/submm telescope which will revolutionize star and planet formation studies after it will be ready in 2010.

- 64 12 m diameter antennas
- 5000 m altitude
- antennas distributed over 10 km across
- total collection area of over 7000 m^2

- $\nu = 30$ - $950\,\mathrm{GHz}$ ($\lambda = 0.35$ - $10\,\mathrm{mm}$)
- angular resolution about 10 milli-arcsec (at 0.5 mm)
- the construction phase will begin in 2003 or 2004
- the full array is likely to become operational by 2010 – the same time when the Next Generation Space Telescope (NGST) will fly

Fig. 45. The Atacama Large Millimeter Array (ALMA)

Subject Index

Θ^1A 190
Θ^1B 190
δ Scuti stars 88, 89, 91
ρ Ophiuchi 116, 121
ρ Ophiuchi 97, 116, 121, 136
θ^1A 177
θ^1B 177
θ^1Ori C 190
θ^1Ori D 190
θ^2A 177
θ^2Ori A 190
θ^2Ori B 190
$^1\theta$ Ori Ic association 123
30 Doradus 26, 136, 209, 216, 220

accretion
- phase 20, 26, 32, 41, 44, 45, 48, 79, 97, 102, 143, 147, 159, 166, 168
- time 24, 48, 49

accretion flow
- geometry of 26, 75

adiabatic phase 15, 20
ambipolar diffusion 18, 113, 129, 179, 208
- timescale 18

B1 122
B5 122
BD+31 643 123, 202
BD+65°1637 106
BD+65°1637 105
binaries
- astrometric 184, 186
- distribution of semimajor axes 174, 175, 179
- eclipsing systems 54, 56, 58, 60, 61, 64, 182
- formation of 13, 60, 73, 112, 166, 172, 180, 182
- frequency of 60, 135, 172, 175, 177, 180, 184, 185, 187, 192, 194
- jets in 177, 179
- massive stars 189
- population of 60, 61, 109, 112, 126
- resolved disks in 141
- rotation axes in 182
- separations of 112, 175
- spectroscopic 61, 62, 73, 88, 184–186
- surveys of 64, 189
- unresolved 64, 65, 172, 173, 201

binary system 180
bipolar cavities 112
birthline 26–29, 32, 42–44, 49, 50, 53, 61, 64, 74, 84, 86, 89, 90, 97, 118, 119, 121, 123, 124, 126, 127
BL 50 91
bolometric corrections 57, 63, 211
Bonnor-Ebert critical mass 16
brown dwarf 39, 41, 54, 72, 73, 121, 206
- lithium test 72, 73
- PPL 15 55

BW3 V38 56

Canuto & Mazzitelli theory of convection 40, 52, 54
centrifugal disk 19, 152
Ceph A 150
Chamaeleon 136, 141, 144, 170
Chamaeleon cloud complexes 75, 97, 112, 120, 129
circumbinary disk 63
circumstellar disks 61, 74, 75, 81, 82, 85, 88, 98, 109, 111, 140, 143, 146, 147,

150, 152, 165–167, 169, 170, 177, 179, 180, 206, 211
circumstellar envelope 152
Class 0 object 17, 99, 118, 123, 136
Class I object 118, 120, 141, 143, 161, 206
CM Dra 55, 57, 58
collapse phase 9, 12, 15, 16, 18, 20, 136, 179, 182
collision/coalescence scenario 51, 189, 195
colour index 57
convection 10, 32, 34, 37, 40, 45, 47, 51, 71
convective phase 44, 71
core collapse 17
Corona Australis complex 64, 136
COROT satellite 94
crossing time 11, 109, 130
CTTS 86, 118, 119, 121, 128, 167, 168, 170, 180

deuterium burning phase 21, 22, 24, 25, 27–29, 40, 43, 44, 46, 47, 79
– duration of the 43, 46
deuterium generated luminosity 22, 44, 45
deuterium main sequence 43
DG Tau B 140
disk accretion 19, 27–30, 32, 74, 75, 82, 97, 104, 109, 155, 157, 161, 168, 180
– shocks 13
disk around early type stars 98, 110, 111
disk debate 109
disk dissipation timescale 167
disk formation 26, 143, 169
disk kinematics 111
disk lifetime 82–84
disk luminosity 49, 74, 75, 83
disk mass 110, 141
disk radius 19
disk rotation 148, 149
disk temperature 141
disk thickness 141
disk wind 104
disk-locking scenario 82, 83
disk-regulated rotation 84
disks inclinations 141, 180
dust 13, 15, 64, 88, 97, 98, 109–111, 137, 140–142, 144, 149, 154, 155, 159, 162, 165–168, 189, 211
– depletion of 49, 97
– temperature of 168
dynamical age 147
dynamical collapse 13–15, 18, 21
dynamical friction 115, 180
dynamical timescale 17, 19, 99, 109, 130, 217

entropy
– evolution of 28, 44, 46, 47
episodic accretion 177
equation of state 11, 16, 36, 37, 51

fragmentation phase 14, 179
FU Orionis 177
fully convective stars 10, 11, 22, 23, 33, 34, 44, 47, 76, 86
fully radiative stars 47

G29.96−0.02 99, 100
G339.88−1.26 112
GG Tau 61, 63
GJ 2609 56
GM Aurigae 140
Gould Belt 129, 137
grains 38, 49
– depletion 49
gravitational collapse 13–15, 129, 155, 189
gravitational contraction phase 24, 25, 32, 44, 48, 91

H_2-jet 156, 161–164
Haro 6–10 177
Haro 6–5B 140
Hayashi phase 47
Hayashi temperature 12
Hayashi tracks 11, 12, 40, 44, 47, 52, 53, 64
HD 123445 188
HD 137059 182
HD 200775 26
HD 34282 26
HD 36981 190
HD 37115 190
HD 37150 190

HD 104237 88, 91
HD 142666 91
HD 35939 91
Herbig Ae/Be stars 26, 32, 35, 44, 64, 75, 88, 105–107, 109–111, 119, 177
– age of 42
– in HR diagram 24, 26, 28, 41, 97
– instability strip of 91
– rotation 81
– variability 88
Herbig-Haro Jets 147, 150, 151, 158, 160, 161
Herbig-Haro object 137, 150, 162, 179
HH30 140, 141, 148, 152
HH 111 146–154, 177
HH 121 177
HH 211 150
HH 212 137, 143, 146–148, 150–153, 157, 161, 162
HH 46 145
HH 47 145, 150
HH 91 150
Hipparcos 26, 28, 93, 121, 128, 137, 184–187, 197, 198
HK Tau/c 140, 141
homologous contraction 11, 46, 47
Hot Cores 99
HP 57 91
HR 5999 88, 91, 119
HR 6000 119
Hyades 61, 71, 81
hydrostatic core phase 15, 20
HII region 216
– ultracompact (UC) 100, 103
HII region 98, 99, 101–104, 136, 137, 208, 209, 215, 220
– ultracompact (UC) 98, 99, 103, 104, 112, 137

IC 348 202, 204, 205, 207
IC 348 122, 135, 200, 202–206
IDS 15189 - 3823 182
infall 15, 19, 21, 26, 29, 32, 49, 51, 97–99, 104, 115, 136, 138, 152, 189
infall reversal 49, 50, 115
infrared excess 32, 65, 75, 82, 136, 167, 169, 177, 211
infrared jets 147, 151
infrared sources 12, 32, 48, 57, 105, 112, 118, 120, 123, 141, 143, 177, 178, 180, 203
initial mass function 53, 66, 115–117, 123, 135, 167, 187, 197, 198, 200, 201, 208–210, 215–218, 220
inside-out collapse model 14, 16, 19, 20
instability strip 89–94
ionization 18, 36–38, 89, 99
– of helium 10
– of hydrogen 10, 12, 38
IP Per 105
IP Per 105
IRAM 04191 118
IRAS 04302 140
IRAS 05413 - 0104 137
IRAS 05491 +0247 179
IRAS 23385+6053 99, 101, 102
IRAS sources 99, 163
IRAS 20126+4104 113
IRAS 20126+4104 111
IRc2 111, 137
isochrones 42, 53, 56, 59–61, 63, 64, 117, 124, 126, 127, 201, 212
isothermal contraction 15
isothermal equation of state 16
isothermal phase 15
isothermal sound speed 13
isothermal sphere 14–16, 18
– Bonnor-Ebert 17

Jeans criterion 14, 17
Jeans length 14, 112, 113
Jeans mass 209
jet
– driving mechanism 147–149, 153, 157, 165
– extragalactic 147
– morphology 150, 164
– shock processes in 150, 151
jet bow-shocks 147, 148, 150, 153
jet formation 146, 161
jet kinematics 147, 151–153
jet-disk connection 157, 158, 161, 162

Kelvin-Helmholtz time 9, 11, 26, 34, 47, 49, 65, 82

L1448 122

L1455 122
L1495E 53, 55
L1527 118
L1640 97
L1641 115, 162
L1688 121
L1689 121
lithium abundance 40, 79
lithium burning 70, 72
lithium depletion 41, 70–74, 129
lithium in rapid rotators 71
lithium line 72, 73, 118, 121, 128, 201
logatropic equation of state 16
LP Ori 190
Lupus region 119, 129, 136, 188

M33 136
M 82 209, 216
M 83 209
Mach disk 151
magnetic braking 80, 82, 179
magnetic critical mass 18
magnetic dissipation 18
magnetic dynamos 167, 205
magnetic fields 12, 18–20, 26, 82, 88, 151, 167
– geometry 82
magnetic pressure 13, 16
magnetic pseudo-disk 19
maser 98–100, 111, 113, 137, 138, 148
mass accretion 76–79, 99, 115, 130
– Bondi-type solution 14
– critical 101, 102, 104
– luminosity due to 13, 34, 75, 78, 101, 172, 180
– rates 13, 14, 16, 17, 20, 22, 24, 26–30, 32, 33, 41, 48–50, 74–76, 78, 79, 82, 91, 97, 99, 101, 102, 104
– scenario 49, 50, 97, 98, 189
– shock 15, 21, 49, 77
mass segregation 108, 109, 115, 217
mass–luminosity relation 54, 56, 172, 218
mass-radius relation 13, 20, 22–26, 28–30
massive star formation 49, 50, 98, 107, 115, 117, 137, 189
massive YSO 98, 102
minimum mass detectable 59
mixing length theory (MLT) 40, 41, 52
molecular clouds 12, 14, 16, 72, 97, 98, 115, 116, 118, 120, 128, 136, 208, 209
– angular momentum 80
– cores 13, 18, 19, 27, 42, 106, 112, 113, 146, 155, 156, 161, 189, 202
– dark clouds 119, 120, 123
– in Orion 97
– in Serpens 116
– lifetime of 129, 130
– line width-density relation 16
– magnetic fields in 18, 26, 129
– morphology 111
– rotation 19, 26
– stability of 17, 18, 27
– star formation efficiency in 12, 114, 128
molecular clumps 98, 99, 130, 156, 157
molecular jets 159, 162
molecules 38, 39
– H_2 dissociation energy 10
Mon OB1 association 123
Mon R2 108

NGC 1333 123
NGC 2024 108
NGC 2070 215, 216
NGC 2264 117, 123
NGC 3603 26, 208–213
NGC 3603 135
NGC 604 136
NGC 6397 217
NGC 6530 117
NGC 6611 26, 97
NGC 6823 91
nonhomologous contraction 44, 47, 48
NU Ori 190

OB association 135, 136, 166, 167, 169, 172, 184, 197, 201
OGLE 56
opacity 10, 12, 36–39, 49, 51, 70, 71
– grains 38, 39, 49
– molecular 38, 89
opacity gap 21
Ophiucus 112, 146, 172, 175

Orion 81, 82, 84, 93, 97, 107, 124, 129, 136, 140, 148, 149, 157, 159, 164, 166, 167, 169, 177, 179, 184, 189, 194
Orion A 137, 155, 157, 159, 161–164
Orion B 137
Orion BN/KL nebula 138
Orion Ib association 151
outflow 34, 51, 80, 82, 87, 99, 102, 114, 115, 126, 130, 145, 147, 148, 150, 155, 157–159, 161–163
– bipolar 32, 113
– cavities 143, 144, 147, 152
– mass 115
– molecular 99, 147, 149, 150, 157, 161, 162, 165
– velocity 115

Par 1605-1 192
Par 1863 191
Par 1863-1 192
Par 1865-1 192
Par 1891 191
Par 1891-1 192
Par 1993-1 192
Par 2074-1 192
Par 2271-1 192
Par 2425-1 192
partially convective stars 47
Per OB2 association 123, 202
Perseus complex 122
photoevaporation 104
planet formation 135, 140, 166, 167, 169, 224
Pleiades 55, 71, 73, 81, 84, 108, 109
polytropic relations 11, 16
post-T Tauri gap 72
post-T Tauri stars 72, 81, 121, 122, 128, 129, 170
PPl 15 73
Prandtl's theory 40
pre–main-sequence phase 211
pre-main-sequence phase 9, 24, 32, 34, 40, 41, 44, 48, 70, 74, 76, 78, 80, 81, 97
Pre-MS luminosity function 218
protostar 12–14, 20, 21, 23, 25, 26, 28, 29, 32–34, 44, 48, 49, 90, 97, 99, 102, 115, 135, 136, 147, 155, 159, 163, 206
– deuterium burning in 32, 44
– isolated 97
– low-mass 10, 22, 137
– mass 9, 19
– massive 98, 99, 102, 104, 137
– maximum radius of 10
– phase 9
– thermal energy 10
protostellar
– accretion 28, 129
– cores 13, 32, 116, 157, 159
– disk 161, 162
– evolution of 21, 32, 117, 157
– jet 137, 146, 155, 157, 161
– luminosity 20
– radius 26, 27, 32
protostellar cloud 14, 15
– collapse of 13, 15, 17, 49
– structure of 15, 16
protostellar phase 33, 35
pulsational properties 88
pulsational variability 87, 88, 91

R CrA 141
R136 26, 220
radiative barrier 24, 33, 44
radiative core 11, 44, 72, 79, 86, 87
radiative track 11, 35, 41, 44, 47, 52, 58
relaxation phase 35, 89
relaxation time 108, 109
residual accretion 51, 74, 78, 80, 81, 126
ROSAT All-Sky-Survey 118, 119, 121, 123, 128, 129, 137, 167
rotation 26, 51, 70, 71, 76, 80, 82–88, 111
– lithium depletion problem 72
– of CTTS 80–82
– of molecular clouds 19, 26
– of PMS stars 71, 80
– of WTTS 82
rotational evolution 83, 84
rotational velocities 80, 83, 85–87
RXJ 15252–3845 182

Sco-Cen 129, 136, 166–170, 175, 184–187
Scorpius 124, 184, 186
sequential star formation 117
Serpens 141

shock-excited nebulosity 137, 147, 157, 179
shocks
– between jets and ambient gas 147
– C-shock 151
– J-shock 151
spherical collapse 49
star formation efficiency 113–116, 129
star-disk interaction 82, 111
stellar mergers 98, 107
stellar winds 32, 74, 114
Strömgren radius 101, 103
Sz 24 172, 173

T associations 119, 120, 125, 136, 167, 172, 184, 197, 200
T Tauri 34, 41, 44, 53, 56, 63, 72, 75, 106, 107, 110, 112, 117, 118, 124, 128, 137, 142, 173, 174, 185, 202–206
– age of 42, 168, 174
– binary system 172, 175, 177, 184, 185
– classical 80–82, 109, 137, 166–168, 205
– classified as 32
– in HR diagram 26, 97, 166, 168
– in Taurus-Auriga 140
– lithium 72, 73
– low mass 34
– mass accretion 76, 172
– mass outflow in 74
– naked 167
– post-T Tauri problem 72, 117, 128, 129
– radii of 10, 42, 80
– rotation 80–82, 84
– variability 88
– weak-line 82, 137, 166–170, 184
Taurus 53, 54, 65, 75, 112, 129, 146, 159, 167, 172, 175, 184
Taurus-Auriga 18, 55, 63, 82, 97, 107, 112, 118, 119, 127, 130, 136, 140, 175, 200, 211
– binary/multiple stars in 112
– low-mass end of the HRD 53, 56
thermal relaxation 35, 41, 47, 48
Trapezium 26, 83, 97, 108, 109, 112, 114, 115, 123, 175, 177, 189, 208, 209
truncated disk 82
truncated IMF 209, 218
TY CrA 60, 61, 64, 182

ultracompact phase 103
ultraviolet excess 82, 167
Upper Centaurus Lupus association 188
Upper Scorpius 121, 127, 197–199, 201, 202
UY Aur 177

V 380 Ori NE 157
V351 Ori 92, 93
V351 Ori 91–93
V372 Ori 190
variability 88, 177, 180, 205
– spectroscopic 88
virial theorem 10
viscous dissipation 75, 141
VV CrA 177, 178

W49 99
WH 349 190
WTTS 72, 118–123, 128, 166–170, 180

X-ray emitting stars 72, 166–169, 184, 185, 201, 203, 205

YY Gem 55, 57, 58

GPSR Compliance

The European Union's (EU) General Product Safety Regulation (GPSR) is a set of rules that requires consumer products to be safe and our obligations to ensure this.

If you have any concerns about our products, you can contact us on ProductSafety@springernature.com

In case Publisher is established outside the EU, the EU authorized representative is:

Springer Nature Customer Service Center GmbH
Europaplatz 3
69115 Heidelberg, Germany

Batch number: 10584688

Printed by Printforce, the Netherlands